100 SAT® MATH TIPS AND HOW TO MASTER THEM NOW!

Written by Charles Gulotta
Illustrated by Trish Dardine

Published by
Mostly Bright Ideas

YOU MEAN I'M
SUPPOSED TO PAY
FOR THIS STUFF?

ISBN: 0-9653263-1-4

Contents

Dedication

This book is dedicated to anyone who's ever experienced math anxiety, SAT panic, or some combination of the two. It's also dedicated to anyone who's ever been pushed around by a bully or held back by self-doubt. You're stronger than any of those things. And, yes, you can even do percents.

Introduction

One million students take the SAT each year. Most score about
500 on math. What is the probability that you will do at least as well?

(A) 0   (B) 20%   (C) $4\pi - 18$ (D) I and III only   (E) Very high, if you just do what I tell you!

Okay, here's the situation. There's math, and there's SAT math. Big difference. I don't care
what grades you've been getting in algebra. I don't care if you've been tutoring Pythagoras's great-
great-great-great-grandson in geometry. I don't care if you can even do cube roots without a calculator.
If you sit down to take the SAT and expect to be given a math test, you're going to get squashed like a
wet gnat in the gutter at a second-grade bowling party.

But wait a minute, you're probably thinking. Isn't the SAT half math? Yes. So isn't it a math test?
Yes, it is, but it's more, and that's where it starts to get dangerous. Up until now, almost every math
test you've ever taken was written or at least planned by your teachers. And your teachers, sweet as
they were, wanted you to do well. They did what they could to help you. They were on your side.

Well, welcome to the real world! The people who write the SAT don't want you to do well.
They don't even like you. Their test is meant to weed out mathematical pinheads such as yourself. And
it's meant to embarrass all of you Einsteins who've been acing math since first grade. Further, it's
designed with such craft and such sinistry that you will take the test, go home still thinking you're a
genius, and then pass out from shock when your scores arrive. (That's the biggest thrill for these
testmakers. They have installed tiny video cameras in the test score envelopes. When you take your
first look at those scores and all the blood drains from your body, they go out for ice cream.)

So what now? Read this book. Use it as a workbook. Fold down the corners of the pages. Write in the
margins. Spit on it. But do something! Open the blinds and let some light in! Don't let them get you!
You're too smart for them. But *you* have to do the work, and it's time to get started.

(I know you don't want to do this, believe me, I understand. Well, tough, you have to. Unless you
want to spend your entire working career saying, "Small, medium, or large?" you'd better get moving
now. So come on. It'll be fun. Sort of.)

What is SAT math all about?

The SAT has sixty math questions: 35 multiple choice (5 answer choices), 15 quantitative comparisons
(4 answer choices), and 10 student-produced responses, or grid-ins (fill-in-the-blank, no answer
choices, nothing but white space, it's all up to you, get the picture?)

Some of the questions are pure arithmetic, others involve algebra, while still others require geometry.
Many of the questions combine two or three of these areas. The math itself is pretty simple, but here's
the hard part: ALL REQUIRE THINKING. Sorry, I hate to have to tell you that, I really do. But you
were going to find out sooner or later.

Thinking. We hate to think, don't we? Wouldn't it be great if we could just sit under a tree and stare at the clouds, then go in for lunch, take a nap, watch a movie, eat dinner, and go to bed? That's what you did when you were three, and it was great, but you don't remember any of it. Now you're approaching adulthood and you have to take the stupid SAT. And to think, when you were a kid you couldn't wait to grow up.

Anyway, back to thinking. Try to understand this, because this is the key to the whole thing. In fact, I'm going to put it in bold letters, so you can find it later.

The key to the whole thing:

Every math question on the SAT can be solved in at least three ways. There's the long route, a middle road, and the direct, bam-you're-there approach. You want the third route. The shortest trip. The one involving the least amount of work.

Why? Several reasons. The SAT has sixty math questions. You have 75 minutes to answer them. That gives you, on average, one minute and fifteen seconds to read the question, solve it, identify the correct answer, mark it on your answer sheet, and move on. So there's a certain element of pressure involved here. If they gave you eight hours to finish the test, or a long weekend, you would get an unbelievably satisfactory score. But the clock will be moving, and more quickly than you can imagine.

The faster you get to the correct answer, the less time you will spend (ooh, write that one down). But also, the less energy you will spend, and that's important because stress and mental exertion can make you extremely tired. The more tired you are, the less focused you will be when you get to the more complicated questions, and the more likely you will be to make careless mistakes. And speaking of careless mistakes, you will keep them to a minimum if you do less work.

Yes, you read that right. The long route to the answer involves more calculations, and the more calculating you do (even with a calculator), the more chances you'll have to make a mistake. Which brings us to...

The second key to the whole thing:

When you're dealing with SAT math, it isn't the math that's hard, it's the SAT. The way the question is asked is much more important than the math it's testing. The language of the SAT is like a brick wall completely surrounding the actual math. Once you learn how to break through the wall, you'll discover a weak, non-threatening little math question cowering, trying to hide from the light. But until you learn the language, you'll continue to think these are hard math questions.

Don't believe me? What if I gave you this math problem to solve:

$$| \heartsuit | \; \mathfrak{R} \; \infty \; \mathfrak{I} \otimes \supseteq \varnothing \propto$$

Oh, I forgot to tell you: this question is from last year's Martian SAT. Does it make a difference that the question is written in Martian? It may, so here, let me translate it into English for you:

$$17 - 3 =$$

Seems like a pretty easy math question, doesn't it? So why couldn't you solve it the first time? Because the math was hidden behind unfamiliar language. That's kind of what happens to many people who take the SAT here on Earth. Most of it is in English (there is a little Martian on the SAT), but the language is used to hinder rather than to clarify. The result, very often, is that the test-takers spend most of their time trying to translate, and have too little time left for doing the math.

Enough philosophy.

Here's an example that will clear up this entire Introduction -- an actual SAT question, changed enough (I hope) to keep away the copyright lawyers:

In February of a particular year, the 23rd of the month falls on a Thursday.
On what day of the week would the 11th fall during that same month?

(A) Wednesday
(B) Thursday
(C) Friday
(D) Saturday
(E) Sunday

As usual, there are at least three approaches. Many people would actually draw out a calendar of the entire month. It certainly seems to be the safest way to solve the question. In fact, you'd almost be guaranteed of getting the correct answer. And it's probably what they're hoping you'll do, because even though you'll get it right, you will have wasted a great deal of time answering question number 3. And that means you'll have that much less time for question 15, and 21, and 26...

The middle road is the one taken by most students. They will say, "Hey, if the 23rd falls on a Thursday, then the 16th also falls on a Thursday (23 minus 7). Now all I have to do is count back 5 more days to the 11th." And that's true.

But think. If the 16th is a Thursday, then isn't the 9th also a Thursday (16 minus 7)? Now all you have to do is count forward two days to Saturday, the 11th. And where are you more likely to make a mistake -- counting forward two days or backward five? The answers are spaced one day apart, so all it would take to arrive at the wrong answer is to be off by one day.

Another thing. Notice which month they picked: February. The weird one. The one with twenty-eight days. That has nothing to do with solving the problem, but a certain percentage of people will have a panic attack: "February! Twenty-eight days. How does that change things? And what if it's leap year? Do I add one to my answer?" And there goes an easy point right down the drain.

This happens to be one of my favorite SAT questions. And that's strange, I know, to have favorite SAT questions. It's kind of like having a favorite poisonous snake, or a favorite oppressive dictator. But after a while (and when it doesn't affect your life much) you learn to admire the trickery inherent in so much of the SAT.

If you read this book and use it the way it was intended, you might even develop that kind of admiration yourself. More important, you will develop the ability to see through these questions for what they really are: bullies who put up a good front, but are really harmless little math questions hiding in the dark behind a wall. They're like those creepy-looking spiders that scared us to death when we were little, causing some well-meaning adult to say, "You know, it's more afraid of you than you are of it." (My first thought was always, "Oh yeah? Then how come the spider is still sitting on my chair and I'm the one standing in the sink?") Anyway, trust me on this one. Those SAT questions are trembling inside. They're terrified, and what they're terrified of is *you*.

So let's get out our flashlights. It's time to climb some walls.

What is math, anyway?

I believe math is nothing more than relationships. It's a lot like the animal world. A gorilla is bigger than a grasshopper. You'll never see a grasshopper bigger than a gorilla (unless you read the supermarket tabloids on a regular basis.) A leopard is faster than a cow. Most fish swim much better than most elephants.

In a somewhat similar way, you began to learn the basics of mathematical relationships even before you went to school. You knew that having two apples was better than one (and three was even better). You knew 10 was more than 5 and a million was bigger than any other number you could imagine.

Eventually you got all the numbers straight in your head and developed ways to judge their relative size, so that now, you instantly recognize that 14,744 is larger than 7,928, even though this may be the first time you have ever seen those two numbers together in the same room.

You learned to add numbers, and soon realized that the result of adding two or more numbers gave you a bigger number. So addition came to represent something that changed the mathematical relationship by creating something larger. Subtracting, on the other hand, did the opposite. If you had seven candy bars and your mother took five away, you had only two left. (It seemed irritating at the time. We had spent all that time learning two apples were better than one, so why were we always having our candy bars confiscated? But by the time we reached high school and took Health class, we realized Mom was doing the right thing.) Subtraction, then, made things smaller.

The following year, we learned that multiplication made things way bigger, and so we knew that five times eight must be bigger than two times eight. Division was a little more mysterious because it used that strange symbol that from the back of the classroom looked like a plus sign. Long division was even more confusing. It put one number under what looked like a carport and the other outside the carport, and then you had to divide, put more numbers on the roof, then multiply, subtract, bring numbers down, and divide again. This was getting to be work, but we eventually conquered long division and found ourselves at the top of the mathematical world. At least for the rest of fifth grade.

Around sixth grade, things began to get out of control.

They started talking about fractions, decimals, common denominators, square roots, and factors. When we began to accept those things as pretty normal, they gave us negative numbers. Then letters instead of numbers. Then letters and numbers. Then adding, subtracting, multiplying, and dividing letters and numbers. Some of the letters had little raised numbers next to them. They told us we could rearrange all these symbols into expressions that could help us figure out how much older Sally was than Billy, or who had more goldfish, Brian or Monica. (As if we really cared. We just wanted to get our candy bars back.)

Next, they showed us pictures of squares and triangles, and for about five minutes we began to feel comfortable again. We already knew what squares and triangles were. We had colored many of them with large crayons in kindergarten, and then at home while showing off for our aunts and uncles. But our comfort was not to last. These squares and triangles were not for coloring. They were just another excuse to add and multiply with letters and numbers. We had to find the perimeter of the rectangle and the area of the triangle, and soon, the circumference of a circle. Our heads were crammed and confused and for many of us, this was where it all came tumbling down. We spent the next two or three years just bluffing our way through math, trying to stay out of the way, sliding by. We never raised our hands in class and always made sure our eyes were looking down whenever the teacher asked a question. We had gotten lost so long ago, we couldn't even remember where we were trying to go. There was no hope of finding our way out of the forest. We would just sit and wait for someone to clear the trees and build condominiums.

And we would've made it, too. Most teachers were perfectly willing to give us a C in math just to get rid of us. Next year's teacher would surely fix the problem. We continued to succeed in this giant slalom of failure, somehow making all the necessary turns, just nicking the poles, but never knocking them over. We were almost down the course.

Then came the SAT.

That miserable concentration of torment and frustration. One small test, one major influence on future events. Why is it so important? Why is it so difficult? Why do I have to take it?

The answers are:
Because.
It isn't.
You don't.

The SAT is important because colleges need a way to gauge your math and language skills in a way that doesn't give anyone an advantage. An "A" average may not mean much because at your high school, maybe almost everyone gets A's, while at another school, almost nobody does. With a standardized test, everyone answers the same questions and is rated against the same scale.

Why does the test seem so difficult? Because it's tricky. It requires you to think. It isn't always what it seems. It can't be trusted, and that's a new experience for you in the test-taking arena.

If you have no intention of ever going to college, don't take the SAT. Save the registration fee and buy yourself a nice tee-shirt. But if there's a chance you might go to college someday, it's probably best to take the SAT now, while the stuff is still fresh in your mind (I make certain assumptions there).

Let's proceed on the premise that you are going to take the SAT, and sometime within the next twelve months. What can you do to improve your chances of scoring high on the math part? A lot. This book will take you on a tour of a hundred typical SAT math questions. It will point out the traps built into each problem, and it will help you ask (and answer) the four key questions you should be asking yourself:

What are they telling me?
Recognizing the information given often requires some translation, and rearranging. This is vital to setting off in the right direction.

What are they asking me?

Frequently, questions are designed so that the goal isn't clear. Or if it is, you end up doing so much work you forget what the question was. Problems that require several steps to solve will take you past a few partial solutions along the way. Usually, these partial solutions will appear as answer choices. It's extremely helpful to restate the question to yourself before marking an answer choice, especially with the more complex problems.

What do they want me to be thinking?

Again, the SAT is designed to lead you in the direction of an incorrect answer. In order to do this, the testmakers try to get you to think about the wrong things. They know you know the math, but if they can disguise the question cleverly enough, you won't know you know how to solve it, and will be tricked into thinking about concepts that aren't going to help you.

What should I be thinking?

When you've done enough practice questions, you will be able to look at most SAT math questions and the math concepts you need to solve the problem will leap out of your mental toolbox -- instantly. And ultimately that's the key: getting those tools out quickly and using them properly. (If you've ever tried to hammer a nail in with the handle of a screwdriver, you know exactly what I'm talking about. And if you haven't, well, go try it.)

Once you've mastered these four elements, you will be on your way to a great SAT math score. The secret is to practice. Do a few SATs, work through this book, then go back and do a few more SATs. You should notice a big difference.

Wait, I almost forgot!

The purpose of an Introduction is to tell you what the book is about, and how it works. (I can't imagine that you're still reading this, but just in case.)

On each of the next 100 pages, you'll find an SAT-type question. The book includes 70 multiple-choice, 20 quantitative comparison, and 10 student-produced response (or grid-in) questions. Each page illustrates a concept, tactic, or math principle that, when understood, will help you handle the SAT with more ease and less anxiety. Obviously, it isn't possible to give you an example of every possible question. The point is not to master the individual problem, but rather to master the kind of thinking necessary to solve these and many other such problems. With a little effort, you'll find yourself wasting less time, getting more right answers, and sweating much less than you used to. I hope you'll also begin to smile a little more. I think this can be fun. Even enjoyable. But then, I really like carrying heavy furniture, so what do I know?

This book has cartoons.

You'll notice that on many of the pages in this book there are cute or amusing little pictures. Some relate directly to the problem, some relate directly to you, and a few, well, I'm not sure what they relate to, but I like them anyway. Just be aware that the SAT doesn't have cartoons. (You probably knew that, but it was worth repeating, I think.)

After the 100th question there are more practice questions, followed by a handy little glossary, and the index. If you look carefully, you'll see that the pages in this book aren't numbered. That's because the questions are numbered and those numbers would not have matched the page numbers, causing great confusion for the printer, who already has enough problems. The result, I feared, would be a lot of incorrectly-numbered pages, and that just wouldn't look good. Especially for a math book.

Just a few more things to keep in mind.

• Unless you're shooting for a perfect score, you don't need to answer every question on the SAT. What you're trying to do is get as many right answers as you can, and leave blank all of the questions you would have gotten wrong. If you worry about leaving blanks, don't. Very often the time you waste on those really complicated questions at the end can be better used by checking your answers on earlier questions. Remember: each answer, no matter how long it takes you, is worth one point.

• The questions presented in this book appear in no particular order. On the actual SAT, questions are arranged according to their alleged level of difficulty. I believe that most of the so-called "harder questions" are not really harder -- they're just more complex. They have more steps, and you have to figure out what those steps are, and the order in which to do them. But the steps themselves are no more difficult than the single step you need to solve question 1.

• The SAT features an assortment of different types of questions: arithmetic, algebra, geometry, word problems, multiple-choice, quantitative comparisons, grid-ins. As you move from one to the next, your brain has to keep shifting gears, because different parts of your brain handle different kinds of thinking. That's one of the reasons this test can be so tiring. I don't know if it helps to know this, but there it is.

• The strangest of all SAT questions has to be the Quantitative Comparison. Here's how it works. They give you two columns, A and B, each containing some value. The value may be just a plain old number, an algebraic expression, the area of a geometric figure, or something even more annoying. Your task is to compare the value under column A with the value under column B. If column A is larger, the answer to that question is A (you would mark "A" on your answer sheet). If column B is greater, the answer is B. If they're equal, the answer would be C. And if it's impossible to tell because they didn't give you enough information, the answer is D. (There's no choice "E.") Questions 51-70 in this book are examples of quantitative comparisons. For added fun, there's another page of them in the back.

• Your math vocabulary is important. Review the words in the Glossary at the end of this book and make sure you understand what they mean. If you still confuse "area" and "perimeter," for example, you need to work on that. The SAT will expose such weaknesses and show you no mercy.

• Have a goal. Do you want a 450 in math? A 500? A 680? Figure out how many right answers you'll need to get, accompanied by a realistic number of wrong answers. Give yourself permission to leave a certain number of answers blank (then you won't panic when you do -- it'll just be part of your plan). Most people are surprised by this, but if you answered just 33 of the 60 math questions correctly, got 16 wrong, and left 11 blank, your score would be at or very near 500.

• You have whatever it takes to get the score you want.

Good luck!

Charles Gulotta

1

**Don't make questions more complicated than they are.
The SAT has already done that for you. Simplify!**

7 (5 + 4) - (3 x 11) =

(A) 23
(B) 30
(C) 63
(D) 96
(E) 2,079

This may appear to be an easy arithmetic problem, and it is. However, as with all SAT math questions, there is a longer way to solve it and a shorter way. And of course, there are traps.

You learned long ago in your mathematical life that $x (y + z) = xy + xz$. This is called the distributive property of multiplication over addition. You have seen so many examples of this rule over the years, that very often your brain automatically performs the operation, even when it isn't necessary. That's precisely what the testmakers are hoping you'll do with this question.

Clearly, the simplest approach would be to add 5 plus 4 and multiply the result by 7 to get 63. Then multiply 3 times 11 and subtract 33 from 63 to get 30 (answer B). But let's look at some of the horrible things that could happen if you choose to distribute.

If you multiply 7 times 5, then 7 times 4, and add the two products, that's three different places where you could make a mistake. Then you have to multiply 3 times 11 and subtract 33 from your first answer. So in this simple little problem, there are five places where a careless error could cost you.

The real math involved here is 63 - 33. Why didn't they just ask it that way? Because the SAT version is designed to trip you up, and lead you toward choosing the wrong answer — or at least wasting a lot of time. They do that by taking a question you could've answered in the fourth grade and mutating it into a form that might cause trouble.

So what's my point? Just this: you probably won't make a mistake multiplying 7 times 4, or adding 35 and 28. But if you *were* going to mess up on those simple matters just one time in your whole life, wouldn't it most likely happen on the SAT? By adding those two numbers inside the parentheses first, then multiplying, you reduce the probability of a mistake by reducing the number of operations necessary. And that's the name of the game here: less work, less time, fewer errors.

Aren't we over-reacting to this problem just a little? This is a mild example, I admit, and maybe you're thinking I've made more of this than it deserves. But if you learn to recognize the trap here, out in the parking lot, you're more likely to see it when we get into the middle of the jungle.

(I know jungles don't have parking lots, but other than that, wasn't it a good analogy?)

Did you notice?	When you multiply 7 times 9 you get 63. Look at answer C. Add 63 and 33, and you come up with 96. Choice D. And if, in a moment of sheer mindlessness, you were to multiply 63 and 33 you'd come up with 2,079. Then you'd mark E on the sheet and move on to question 2, completely unaware of your careless mistake. Don't pick an answer just because it's one of the choices.

2

All questions concerning average have three parts: the number of things, their sum, and their average. Ignore the phrase "arithmetic mean."

The average (arithmetic mean) of three numbers is 31. If one of the numbers is 14, what is the sum of the other two?

(A) 17
(B) 45
(C) 54
(D) 79
(E) 93

Every time you see an "average" question, begin with the formula: $\dfrac{x+y+z}{3} = $ **Average**

If there are four numbers being averaged, put a 4 in the denominator, if five, use a 5, etc. But always begin with this basic formula. The average question has three parts: the number of things, the sum of those things, and their average. The SAT will give you two of them, and ask you for the third. They will also throw in that phrase *arithmetic mean* every time, either to help you or confuse you (which do you think?)

In the above example, they've given you the average, 31. That's one part. They've given you the number of things, 3. That's the second part. At this point, they could ask you for the sum of the three numbers — that would be the third part. But instead, they've given you one of the numbers and want the sum of the other two. So once we've found the sum of all three, we just subtract 14 to arrive at the answer.

Let's take the formula and plug in what we have: $\dfrac{14+y+z}{3} = 31$

This is what they told us. What are they asking for? The sum of $y + z$. So we have to get the sum $(y + z)$ on one side of the equal sign, and the numbers on the other. The first thing we'll need to do is multiply both sides of the equation by 3 to clear the fraction. We then get:

$14 + y + z = 93$

When we subtract 14 from both sides, we're left with the answer:

$y + z = 79$ (Answer **D**)

Did you notice?

As usual, incorrect solutions arrived at through common mistakes are found among the answers. If you subtracted 31 minus 14, you'd get 17 (A). If you added 31 and 14, you'd get 45 (B). If you simply multiplied 3 times 31, you'd get 93 (E). Don't fall into these traps. Average questions are easy. No need to ever lose the point!

3 Use "what if" to quickly arrive at a good approximation. Don't fool with difficult numbers when the easy ones will get you close to the answer.

A beacon rotates so that a stationary observer sees the light "flash" every 29 seconds. If the beacon maintains a constant rate of rotation, approximately how many times will the light flash in one hour?

(A) 36
(B) 60
(C) 124
(D) 2,400
(E) 3,600

First things first: there are 60 seconds in a minute, 60 minutes in an hour. Therefore, there are 3,600 seconds in an hour. Remember this now and in the future. Apparently the SAT people have found that a good number of students buzz out during the test and think there are 60 seconds in an hour. (Very likely that's the way it seems during the SAT, but it just isn't so, and this misjudgment can affect your choice of answers in an extremely negative way.)

Okay, the question tells us that a beacon flashes every 29 seconds. We could divide 29 into 3,600 (the number of seconds in an hour) to get at the exact answer. And with a calculator that would be quick and easy. But wait! Let's practice a technique that will rescue us someday when the numbers aren't so small, the question isn't so straightforward, and the batteries in our calculator are dead. Let's use "what if" to very quickly get close to the correct answer.

What if... the light flashed every <u>30</u> seconds. Then we could say it flashes two times a minute, every minute, for sixty minutes. That would be 120 flashes. Since the question says "approximately," and since we only changed the information slightly (from 29 seconds to 30), we can safely assume that 124 (C) is correct. Which it is. (Notice also that the answers are somewhat spread apart, so approximating is a safe tactic. You can be off a little and still pick the right answer.)

When is it safe to do this? Use this technique whenever the information provided is painfully close to the number you would wish it to be, the question calls for an approximate solution, and the answer choices are not terribly close together. For example, if they tell you something is 99 feet long, or occurs every 23 hours, or has an angle of 89 degrees — try changing the number slightly and you'll probably be able to get an approximate answer very quickly.

Did you notice?	Which answers make sense? If the light flashed once a minute, 60 would be the correct answer. But in fact it flashes more frequently than that, so A and B can be eliminated. It would have to flash every second to hit 3,600 in an hour (E), and every second and a half to reach 2,400. The only answer that could possibly be correct is the one that is: C.

4 You don't necessarily have to know what every variable equals in order to answer the question.

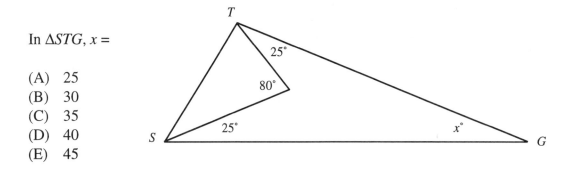

In △STG, x =

(A) 25
(B) 30
(C) 35
(D) 40
(E) 45

Is this problem solvable? At first it may not appear as though they've given you enough information. If you don't know what either angle S or angle T equals, how can you possibly know what angle G equals?

Obviously, you can, or one of the answers would have said, "Cannot be determined from the information given," or something to that effect. This problem can be solved, and by *you*. Let's start by deciding which tool to pull out of your mental toolbox. What priceless gem do you have stored in there that might come in handy? How about the one that says all triangles have 180 degrees? Doesn't that seem to fit the need here?

It's very likely that's where you started. And when that didn't seem to get you anywhere, you probably thought about extending some lines off the small triangle to form even smaller triangles. And when that proved to be a dead end, you considered having plastic surgery, buying a fake driver's license from someone named Eddie, and starting life over in the next state.

Take a good look at the illustration. And as you do, remember this: if you know two angles of any triangle, you can determine the third angle. All you do is add together the measures of the two known angles and subtract that sum from 180. If you knew the <u>sum</u> of angle S and angle T, you could find angle G, or x.

See that?

Since one of the angles of the smaller triangle equals 80 degrees, the other two angles must add up to 100 degrees. When you add 100 degrees to the two 25 degree angles, you see that the <u>sum</u> of all the angles that make up angles S and T in the larger triangle equals 150 degrees (100 + 25 + 25). And if angle S plus angle T is 150 degrees, angle G, or x, must equal 30 degrees.

Answer **B**. Well done!

Since the sum of these two angles is 100 degrees...

...the sum of these two must be 100 + 25 + 25, or 150.

Did you notice?

We never did find out what those two angles in the smaller triangle actually equal individually. But in order to answer this particular question, it really doesn't matter what they equal. What matters is what their <u>sum</u> equals, and that we found with no problem. (We don't know everything, but we know enough!)

5

With two equations, each containing two variables, combine the equations to get rid of one of the variables.

If $v = 2w - 7$ and $3v = w + 4$, what is the value of v?

(A) 3 (B) 5 (C) 7 (D) 12 (E) 15

This may seem impossible to solve, because in each equation more is unknown than is known. The secret is that when you *combine* the two equations, their relationship reveals solid information. You can then implant that information back into the equations to reveal even more.

We're told that $v = 2w - 7$ and $3v = w + 4$. They want to know what v equals, so our goal is to get rid of w. The key to this whole thing is the fact that we can subtract one equation from the other in order to cancel out one of the variables (or add them if that accomplishes what we want.) But before we add or subtract, we need to adjust one of the equations so the unwanted variables are equal to each other in each equation. Look:

If we multiply the second equation by 2, there will be a $2w$ in each equation. Then subtracting them will cancel out the w completely.

$v = 2w - 7 \qquad 3v = w + 4$

$(\times 2)$

$6v = 2w + 8$

$v \ = 2w - 7$

Now when we subtract these two equations we get:

$$5v = 15$$

$$v \ = 3$$

If we wanted to find w, we could plug $v = 3$ back into either equation. For example:

$$3 \ = 2w - 7$$

$$10 = 2w$$

$$5 \ = w$$

Wow!
The really amazing thing about these kinds of equations is that each one can be plotted to form a line on a graph. The place where the two equations we just worked with intersect is the point (3,5), where $v = 3$ and $w = 5$. So solving the two equations means finding the values for v and w which make both equations true. And finding the place where the two lines meet on the graph.

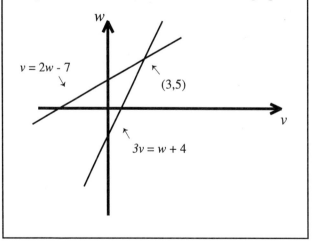

$v = 2w - 7$

(3,5)

$3v = w + 4$

Did you notice?

You can also get rid of the v's and solve for w by multiplying the equation on the left by 3. Then when you subtract the two equations, the $3v$'s drop out and all you have is w's and numbers. Solve for w, then plug that value into one of the original equations and find v. Believe it or not, there are other ways to solve this kind of question. But let's discuss it some other time.

6 Probability, fractions, decimals, and percents. They're all the same idea: comparing a part of something to the whole thing.

A large box contains exactly 24 bowling balls. All of the bowling balls are either black, yellow, blue, or red. The probability of selecting a red ball at random from the box is 1/4, and the probability of selecting a black ball at random is 1/6. What is the sum of the yellow and blue bowling balls in the box?

(A) 8
(B) 12
(C) 14
(D) 16
(E) 18

DOES THIS ONE COME IN PEACH?

Let's begin at the beginning. If the probability of randomly taking a red ball out of the box is 1/4, that means one-fourth of the balls in the box are red. Since one-fourth of 24 is 6, we can comfortably assume there are 6 red bowling balls in that box. In a very similar way, if the probability of pulling out a black ball is 1/6, then one-sixth of the bowling balls are black. And because one-sixth of 24 is 4, we have now determined that there are exactly 4 black bowling balls in the box.

If there are 6 red and 4 black balls, that means 10 out of the 24 are either red or black. Which also means that 14 of the 24 are either yellow or blue. So the correct answer is **C**.

Here's a mini-lesson in probability. It's all fractions, it really is. That doesn't thrill you? A fraction is just a comparison between a piece and the whole thing. (Relationships, remember, from the Introduction? You didn't read the Introduction? Well, go read it now, there's some pretty good stuff in there.)

Let's think about pizza. I was just thinking about pizza anyway, and figured we might as well work it in. If there are eight slices in the pizza, and you're going to eat one-fourth of them, your share is two slices, because one out of every four means two out of eight.

Probability is just the same thing. Put the whole thing on the bottom of the fraction. In the ridiculous bowling ball example, there are 24 balls in the box, so we put 24 in the denominator. Then, because one-fourth of the bowling balls are red, and one-fourth of 24 is 6, we can say that the fraction representing the red balls is 6/24 (or 3/12, or 1/4— they're all the same.) Probability, then, is that fraction expressed in its lowest terms. The red balls are 6 out of 24, so we say the probability of choosing a red one on the first try is 1/4.

Did you notice?	Most of the information given is expressed in terms of probability, yet the question is asking for the number of yellow and blue bowling balls. In order to make the mental shift from probability to number of objects, you have to understand the connection between them. Also, the actual number of yellow balls and blue balls is unknowable -- and unimportant. It's the sum that matters.

7

Don't just solve for the variable -- answer the question. (Part I)

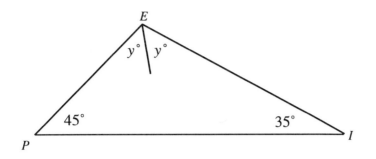

In $\triangle PEI$ above, what is the value of y?

(A) 30
(B) 40
(C) 50
(D) 100
(E) 180

This is a classic SAT question, for a couple of reasons. First of all, it tests one of the basic concepts of geometry, namely that all triangles, no matter which corner of the universe you're standing in, have three angles that add up to 180 degrees. (The only time this changes is when you happen to be traveling at close to the speed of light, because then light bends, you can see around corners, and your mass approaches infinity. What I usually find, though, when my mass approaches infinity is that I need to get out in a hurry and buy some new underwear and the SAT becomes relatively unimportant.)

Where the heck were we? Oh yeah, triangles. The angles add up to 180. And there are three of them. This means that when you know two of the angles, you can always figure out the third one. All you have to do is add the two you know and subtract the result from 180. We've talked about this already, because it sounds familiar, even to me. What you have left is the measure of the third angle.

In the problem you're patiently waiting to solve, the two known angles are 45 and 35 degrees. Their sum is 80 degrees. That means the third angle equals 100. So the answer must be D, right?

Not so fast!

What are they asking you? Do they want the measure of the third angle, or do they want to know what y equals? They want y. Since the third angle equals $2y$ (or $y + y$), we have to cut our answer in half to know what y equals. The answer is not 100, which is what they're hoping you'll think, but rather 50, which is what I'm hoping you'll think.

The correct answer is **C**.

Did you notice?	Your natural tendency in this problem is to solve for the third angle -- that's what your math teacher always asked for. But remember: this is the SAT, and many of the traps are designed around those natural tendencies. When you've figured out what something equals, make sure that's what they're asking for. If they want x, give them x. If they want the largest angle, give them the largest angle. And so on.

8 Sometimes they give you information in different forms and and in different places. It's your job to put it together.

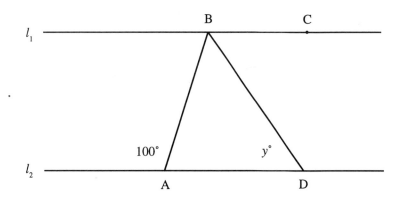

Lines l_1 and l_2 in the figure above are parallel. Line segment *DB* bisects $\angle ABC$. What is the value of y?

(A) 45   (B) 50   (C) 60   (D) 80   (E) 100

As with so many SAT geometry questions, there are several ways to tackle this one. Because it involves parallel lines and a triangle, you'll have to draw from your knowledge of both — and put them together appropriately.

Notice first that they give you the 100° angle in the diagram. What they're hoping you'll do is jump on the fact that $\angle BAD$ is 80°, because those two angles add up to 180°, which is true but will leave you with no place to go from there. Instead, work with what you know about parallel lines. If that given angle is 100° then $\angle ABC$ is also 100°. Now remember that segment *DB* bisects $\angle ABC$. So it cuts it into two 50° angles. And if $\angle CBD$ is 50°, then thanks to the very same rule we just used about parallel lines, so is $\angle BDA$, which is what we're looking for: $y = 50$ (**B**).

Another approach. You can also use the fact that the three angles of the triangle must add up to 180°. Once you've determined that $\angle BAD$ is 80° and $\angle ABD$ is 50°, it's just one more step to realize that $\angle BDA$ must also be 50°

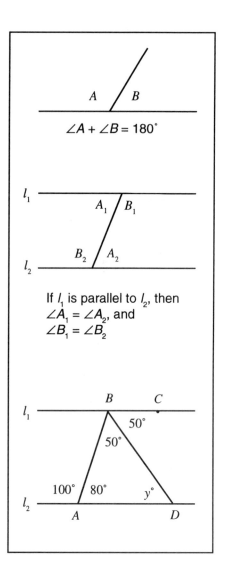

Did you notice? They tell you the value of one angle in the diagram, then they indirectly tell you the value of the others in the problem. I say indirectly because you have to figure out that $\angle ABC$ equals 100°, and you must interpret the fact that *DB* bisects that angle into two 50° angles. They will always provide the necessary information, but they will make you work for it.

9

If they want the area of a circle, you want the radius. Your task is to find the connection between what they've told you and the radius of that circle.

A 6 x 8 rectangle is inscribed inside a circle.
What is the area of the circle?

(A) 6π　(B) 8π　(C) 10π　(D) 25π　(E) 50π

"Are you kidding me?" I can hear you from here. Really, I can hear you. You're thinking, there's no way anybody, even someone who actually likes math, could possibly figure out the area of a circle just by knowing the size of the rectangle inside it. Well, of course they can. And so can you. Start by drawing a picture of the information given.

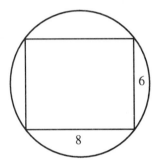

Good. Now what's the connection between the rectangle and the radius of the circle? I don't know either, so let's try something. (Doing something is usually better than doing nothing.) Let's draw the diagonal of the rectangle. This gives us another shape we know how to deal with: right triangles.

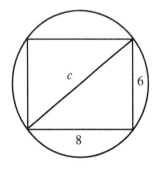

Nice diagonal! Using our old friend, the Pythagorean Theorem, we can now find the length of that diagonal. Why would we want to do that? Because *the diagonal of a rectangle inscribed in a circle is the diameter of that circle!* Are you as excited as I am?

$a^2 + b^2 = c^2$

$6^2 + 8^2 = c^2$

$100 \quad = c^2$

$10 \quad = c$

If the diameter of the circle is 10, then the radius must be 5. Now all we have to do is plug 5 into the formula for area of a circle, and we're done. And miraculously, the formula is handed to us, free of charge, right on the SAT!

$$\text{Area} = \pi r^2$$

$$= \pi(5^2)$$

$$= 25\pi \text{ (Answer D)}$$

| **Did you notice?** | Several things. First of all, they don't give you a diagram. Practice drawing circles, rectangles, and triangles -- it will be a great tool. Second, they give you just enough information. Third, they assume you know what the word *inscribed* means. And fourth, you need to know that rule: *the diagonal of a rectangle inscribed in a circle is the diameter of the circle.* |

10 When adding or subtracting series of sums, find the pattern that will help you avoid endless calculations.

If the sum of the integers from 101 to 200, inclusive, is subtracted from the sum of the integers from 201 to 300, inclusive, what is the result?

(A) 0
(B) 100
(C) 500
(D) 1,000
(E) 10,000

Don't waste precious time. You have an hour and fifteen minutes for all of the math sections on the SAT. You could easily use up half of that time on this question. Needless to say, that would not be a brilliant strategy, and your score would reflect your faulty judgment.

Adding more than a few numbers is never necessary on this test! There's always a faster way. How do we find it? By first finding the pattern. Notice they aren't asking you to find the sum of the integers from 101 to 200. Nor do they want the sum of the integers from 201 to 300. They're asking for the result when you *subtract* the two sums.

Let's set it up in mathematical form. The sum of the integers from 101 to 200 looks like this:

201 + 202 + 203 + ... + 300 (The three dots represent all the integers between 203 and 300.)

The sum of the integers from 101 to 200 can be expressed like this:

101 + 102 + 103 + ... + 200

Do you see that each number in the top sum has a corresponding number in the bottom sum that is exactly 100 less? For example, match 201 with 101, 202 with 102, and so on. Now instead of adding the two lists of numbers and subtracting their sums, let's subtract each pair of numbers. This approach may seem like more work, but it isn't because it will help us quickly identify a pattern.

201 + 202 + 203 + ... + 300
- 101 + 102 + 103 + ... + 200
_____

100 + 100 + 100 + ... + 100

For each pair of numbers subtracted (201-101, 202-102, etc.), the result is always 100. And since there are 100 pairs of numbers, the answer is 100 x 100, or 10,000 (**E**).

| **Did you notice?** | Don't blow a math question because you don't know what a word means. A strong vocabulary is essential to answering this and many other SAT questions. Do you know what *sum* means? How about *integer*? *Inclusive*? If you're not sure (and even if you only think you are), look them up in the glossary in the back of this book. |

Earnings equal the hourly rate times the number of hours worked.
If the information contains letters, and it's confusing, plug in numbers.

Franklin earns *m* dollars per hour for *n* hours. The next day he gets a raise to 3*p* dollars per hour for *n* hours. In terms of *m*, *n*, and *p*, how many dollars did he earn in the two days?

I SURE HOPE MY ACCOUNTANT KNOWS THE ALPHABET.

(A) $2n(m + 3p)$

(B) $m(n + 3p)$

(C) $p(3m + n)$

(D) $n(m + 3p)$

(E) $3m(n + p)$

The real question is, does Franklin know how much he's earned? And the answer is, not if they pay him in letters. How do you deal with this stuff? Try putting yourself into the question. I find that when I personalize it, my thinking becomes a little clearer. Especially when earnings are involved.

If you were making 10 dollars per hour and you worked for 3 hours, how much would you expect to receive? Right, 30 dollars. If you were earning 15 dollars per hour and you worked 4 hours, how much then? Absolutely correct, 60 dollars. (See how sharp your multiplication skills become when we're talking about *your* money?) So the rule is: to calculate total earnings, multiply the hourly rate times the number of hours worked. The formula isn't hard to remember — you just did it without thinking!

Okay, back to the SAT problem. If Franklin earns *m* dollars per hour and he works *n* hours, how much does he earn that day? The answer is *mn* dollars (*m*, the hourly rate, times *n*, the number of hours worked). Are you with me? At the end of day 1, Franklin has *mn* dollars stashed in his wallet. Okay, the next day his boss, recognizing the incredible job Franklin did, raises his hourly rate to 3*p* dollars per hour. (Is 3*p* really more than *m*? Only Franklin and his boss know for sure.) He's still going to work *n* hours. So on day 2, he should be paid 3*p* times *n*. (That's 3*p* dollars per hour times *n*, the number of hours worked).

How much does Franklin earn for the two days?

$$m(n) + 3p(n)$$

or

$$(m + 3p)n$$

If you just flip this answer around, you'll see that it matches answer **D**.
Now for ten bonus points, complete Franklin's 1996 federal tax return.

| **Did you notice?** | The answers all look alike -- because they're expressed in those weird letters. The fact is, I could read that question a hundred times and never get any closer to the answer. My brain doesn't calculate in letters, does yours? It's like trying to inhale mashed potatoes. Our lungs were meant to breathe in air, and our brains were meant to calculate numbers. So plug in some numbers before we suffocate! |

If the question asks for the circumference of a circle, find the radius, then plug it into the formula.

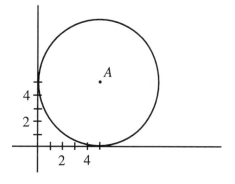

What is the circumference of circle *A* above?

(A) 5π
(B) 8π
(C) 10π
(D) 16π
(E) 25π

This is a very simple problem, so let's not blow it! If they ask for the circumference, you need the radius. Now remember, this is the SAT. They're not going to just hand you everything you need. They've given you all the formulas, and that's about as nice as they're going to get.

First, in the diagram above, "circle *A*" does not refer to that little dot in the middle of the circle -- it refers to the whole circle. We name circles by their midpoints. It's similar, in a way, to how we name tropical storms, but not similar enough to make it worth mentioning here.

Second, circle *A* is sitting on the *x*-axis and is backed right up to the *y*-axis of a graph. Since there is no other information given, **that must be how we're going to find the radius!** So let's get to it.

Actually, there's not much more to do than count. The trick here is to count correctly. The midpoint of the circle lines up with 5 on the *x*-axis, and 5 on the *y*-axis. So no matter which way you go, the radius is 5. Notice they don't number the axes to show the 5. You have to be smart enough to count one line past the 4.

Now all we need is the right formula. The circumference of any circle is found with:

$C = 2\pi r$, where *r* is the radius.

$= 2\pi(5)$

$= 10\pi$ (Answer **C**)

Did you notice?

They give you the formulas, but they don't tell you when to use them -- that's your job. If you use the wrong formula, you will come up with the wrong answer. And there's an excellent chance your wrong answer will be one of the choices. For example, if you used πr^2 by mistake, you'd get 25π. If you counted wrong, you'd get 8π. If you counted wrong *and* used the wrong formula, you'd get 16π.

13

Very often, adding the percents in a word problem will give you a nice round number to work with.

Maria has $20. She owes her brother, Tom, 13 percent of this amount, and her sister, Theresa, another 27 percent. After Maria pays up, how much of the original $20 will she have left?

(A)   8.00
(B)   12.00
(C)   14.60
(D)   16.00
(E)   17.40

ANY MORE SIBLINGS
AND I'D BE BROKE!

Do you see the trap in this problem? They're hoping you'll find 13 percent of 20, then 27 percent of 20, add the two results, and subtract from 20. And you could do that, but it's too much work that will take too much time and expose you to too many possibilities for careless mistakes. Let's play it safe, get the right answer, and move on.

If Maria owes one person 13 percent of her money, and another person 27 percent of her money, then she owes a total of 40 percent of her money to other people. Setting it up mathematically, we get:

$$(.13 + .27) \times 20 \quad = \textbf{ number of dollars Maria owes}$$

$$.40 \times 20 \quad =$$

$$8.00 \quad =$$

So what's the answer? Well, it isn't (A), although that's just what the SAT people are hoping you'll think. The question is asking how much Maria will have <u>left</u> after she pays her debts. She owes 8 dollars. Out of the original 20, she will have 12 dollars left (**B**). Can you believe how sneaky?

If you prefer keeping the percent in fraction form, you would end up multiplying 20 by 40/100. Then you'd get 800/100, which reduces to 8.

Either way, you want to add the percents. And you can use this technique with as many percents as you're given. (In other words, if Maria suddenly remembers she owes her sister-in-law 12 percent and her cousin 26 percent, she now owes 13 + 27 + 12 + 26 percent, or 78 percent.) But remember: you can only do this if the percents are all being applied to the same number, which in this case is 20.

Did you notice?	You can save yourself even more time and work on this problem. Once you've determined that Maria owes 40 percent, you can then conclude that she <u>keeps</u> 60 percent. And since we need to learn how much she has left, all we have to do is find 60 percent of 20 (in other words, .60 x 20, or 60/100 x 20), and we have our answer (12.00) in one step.

14 If you're presented with a cylinder and some liquid, you'll probably need to find the volume of the cylinder. Just use the formula they've given you!

height = 10 in.

radius = 4 in.

Beginning at 1 p.m., the cylinder above is filled with pineapple juice at a constant rate of 16π in^3 per hour. What time will it be when the juice reaches the very top of the cyclinder?

(A) 6 p.m. (B) 8 p.m. (C) 10 p.m. (D) 11 p.m. (E) 12 a.m.

Don't waste time trying to figure out why anyone would spend hours filling a cyclinder with juice. We all do strange things. If filling a cylinder with juice were the strangest thing I'd ever done, I'd be free to run for public office. Let's just solve the thing.

We're dealing with the volume of a cylinder. WHAT'S THE FORMULA? Relax, they give you all the formulas, remember? It's right there with the others, on the first page of each math section of the SAT.

$$V = \pi r^2 \times h$$

Well, what in the world does *that* mean? It means that to find the volume of a cylinder, you multiply π (pi) times the square of the radius, then multiply that whole thing by the height. What you'll end up with is some amount expressed in cubic feet, or cubic meters, or in this case, cubic inches. Let's plug in what we know:

$$V = \pi(4^2) \times 10$$

$$= 16\pi \times 10$$

$$= 160\pi$$

So the volume of this cylinder is 160π cubic inches, or 160π in^3. If the juice is being added at the rate of 16π in.^{3} per hour, it's going to take 10 hours to fill the cylinder ($160 \div 16$). The juice will be right at the top at 11 p.m. (**D**). (Remember, 10 hours doesn't necessarily mean 10 o'clock. And if you were planning to show up at 12 a.m., bring your mop.)

|  **Did you notice?** | The volume of a cylinder is the area of the circular bottom (πr^2) multiplied by the height of the cyclinder. So think of the cylinder as a stack of pancakes. To find the volume of the stack, first find the area of the bottom pancake. Then multiply that area by the height (the number of pancakes). The result is the volume. (Pineapple juice and pancakes -- well wasn't this a yummy little problem.) |

15

Use logic to eliminate wrong answers. Very often, the correct answer is the only one left.

The Hampton Silverware Company's old machine makes 300 spoons per hour. The new model produces 500 spoons per hour. If they start at the same time, how many <u>minutes</u> will it take the two machines to make a combined total of 1,000 spoons?

(A) 40   (B) 50   (C) 75   (D) 150   (E) 180

What do they want you to do? They want you to sit there and stare for about two minutes. Then they want you to try a few equations, draw a picture of some spoons, begin weeping, regain your composure, decide to leave this one blank, and move on to the next question. But you're going to solve it. You have two approaches:

Here's the algebra approach:

The output for each machine can be expressed as *rate x time*. The number of spoons per minute times the number of minutes will tell us how many spoons the machine makes. Since both machines will be starting and stopping at the same time, we know the *time* part of the equation will be the same for each machine. Let's call that *t*. The rate for the older machine is 300 per hour, but since we need the answer to be in minutes, we'll change that to 300 per 60, or 300/60. The output for the older machine (*rate x time*), then, would be 300*t*/60. In the same way, the output for the newer machine would be 500*t*/60. The two outputs together equal 1,000. When we solve for *t*, we'll know how many minutes it would take for the two machines, running at the same time, but at their own rates, to produce 1,000 spoons.

300*t*/60 + 500*t*/60 = 1,000

Multiply by 60 to clear the fractions:

300*t* + 500*t* = 60,000

800*t* = 60,000

t = **75**

Here's the logic approach:

One machine makes 300 spoons in 60 minutes. The other machine makes 500 spoons in 60 minutes. Running at the same time, then, the two machines would make 800 spoons in 60 minutes. But we need 1,000 spoons. So the answer must be more than 60, and we can eliminate A and B. Since the machines running together produce 800 spoons in one hour, they would produce 1,600 spoons in two hours, or 120 minutes. But that's too many spoons. Therefore, the answer must be greater than 60, but less than 120. The only answer that meets those requirements is 75 (**C**).

Did you notice?

The logic side is much shorter than the algebra side. That's no accident. Very often, you can figure out the correct answer using logic -- and do so much more quickly than if you had taken the long, drawn-out, mathematical approach. Is that cheating? No, it's thinking (which is what the SAT is really testing, remember?)

16

If the question uses the phrase "MUST be true," skim the answers and find the one related to a mathematical law.

If s and t are integers, $s \neq 0$, and $-s = t$, which of the following statements MUST be true?

(A) $s > t$
(B) $t > s$
(C) $s + t > 0$
(D) $s - t = 0$
(E) $st < 0$

This seems to be a confusing question, and therefore it is. What makes it confusing? A couple of things. First, it asks your brain to think about a certain relationship between two numbers, but the numbers are expressed as letters. Your brain does not care for this.

Second, they tell you s cannot equal 0, but what about t? Third, they introduce the already-confusing negative sign by pairing it with the equally-confusing letter s. Fourth, they use that word MUST in all uppercase letters and it's very scary.

Finally, the answers make you want to go out and drown a mathematician. They all look like possibilities, so your tendency may be to try out each one by plugging in numbers. Which is precisely what our lovely friends in SAT-land are hoping for. This is a good, old-fashioned time waster.

So what should you do? Take a step back.

If $-s = t$, then s and t are the same integers with opposite signs. If s is positive, t is negative. If s is negative, t is positive.

Answers A and B both could be true, but neither one MUST be true. Remember, s and t are the same integer with opposite signs. Either one could be the positive number.

Answers C and D are both close to being true. If either had said $s + t = 0$, that would be correct, because two equal integers with opposite signs ($-7 + 7$, $-25 + 25$) will always add up to 0.

Which leaves answer E. If s and t have opposite signs, then one must be positive and the other must be negative (it doesn't matter which is which). When you multiply a positive integer times a negative integer, you will always get a negative integer. This is a mathematical law. So the expression that must be correct is found in answer E.

And don't forget: the phrase *MUST be true* doesn't mean for one possibility, or occasionally true. It means for every possibility, and always true.

Did you notice?

The correct answer is E. That means if you decided to use the trail-and-error method (plugging in numbers to see what happens, you will have to go through all the choices before you hit the right one. You might gain a point, but you'd lose valuable time. The solution to this situation is to take that step back. But if you must use trial-and-error, think about trying the last answer first.

17

They're trying to confuse you: all the information is presented, but in an illogical order. Go directly to the hard facts.

In four years, Meaghan will be twice as old as Allison was three years ago. Allison is eleven years old today. How old is Meaghan?

(A) 12
(B) 14
(C) 16
(D) 24
(E) 30

This question could be much clearer if the facts were given in a more logical sequence. For example:

Allison is eleven years old. Three years ago, Allison was exactly half the age Meaghan will be four years from now. How old is Meaghan today?

Still confusing maybe, but much easier to deal with. In the original question, we're told about the relationship between Meaghan's future age and Allison's past age, THEN we're told what Allison's actual age is. In the second, kindler, gentler version, we're told immediately how old Allison is now. Then we're eased into figuring out how old Allison was three years ago, without anything about Meaghan to confuse the issue. Then we move on to Meaghan and stay with her for the rest of the question. It's clearer. Which is why the SAT will never ask a question like that. They're the same math question — but only the first one is an SAT question.

How do you deal with this confusing language? Go directly to the facts. How old anybody was three years ago or will be four years from now does you no good until you know something definite. Allison is eleven. Now stick with Allison: three years ago, she was eight. Okay, now hold onto that and let's move to Meaghan. She will be twice as old as Allison was — in other words, twice as old as eight. So Meaghan will be sixteen. When? In four years. If Meaghan will be sixteen in four years, she must be twelve now (answer **D**). Do you see how the first piece of information given was used in the last step? And the last piece of information given was what we needed first.

For all you algebra fans, here's how to set it up. Let *M* equal Meaghan's age now, *A* Allison's age now. Then, according to the information given:

$$M + 4 = 2 (A - 3)$$

$$M + 4 = 2 (11 - 3)$$

$$M + 4 = 16$$

$$M = 12$$

Did you notice?

Meaghan will be sixteen four years from now. That's not what they're asking, but it sure is one of the answer choices. Also, did you notice that all the numbers in the problem were written out as words, but the answers are presented as numerals? Why do you suppose they do that?

18

When looking for consecutive integers (odd, even, or just consecutive), plug in numbers or set up an algebraic expression.

If a, b, c, and d are consecutive odd integers, and $a < b < c < d$, then $a + b$ is how much smaller than $c + d$?

(A) 8   (B) 9   (C) 10   (D) 11   (E) 12

We're told that there are four consecutive odd integers. That means each one is 2 more than the one before it. For example, 1-3-5-7 would work nicely.

Using these numbers, $a + b$ would equal $1 + 3$, or 4, and $c + d$ would equal $5 + 7$, or 12. Since 4 is 8 less than 12, the answer would be A.

Only trouble is, I don't usually trust one example, so I'd pick another set of numbers and check it again. Like 7-9-11-13. Now $a + b$ would equal $7 + 9$, or 16, and $c + d$ would equal $11 + 13$, or 24. Again, 16 is 8 less than 24, so having found an answer that worked two out of two times, I would go with choice A and would lose no sleep tonight over my selection.

But what if the numbers were a lot bigger and there were more of them? Or what if I just had nothing better to do and wanted to set up an algebraic expression? How would I go about accomplishing such a thing? Probably like this:

The smallest number is a, so let's call it... a.
The next odd integer, b, would equal $a + 2$.
So $a + b$ would equal $a + (a + 2)$.
Proceeding along this incredibly sound line of reasoning,
c would equal $a + 4$, and d would equal $a + 6$. Are you still with me?

Back to the question: $a + b$ is how much smaller than $c + d$?
Well, if all that stuff we just said is true, and you agreed it was, then:

$(c + d) - (a + b) = [(a + 4) + (a + 6)] - [a + (a + 2)] =$

$[2a + 10] - [2a + 2] =$

$10 - 2 =$

8 (Answer A again.)

Did you notice? | There are four variables and they each appear three times in the question. That's twelve variables in one rather short question. Try to notice what the appearance of this question does to you. Is it intimidating? Confusing? It's meant to be those things. But remind yourself that, with all its letters and *less than* signs and *plus* signs, this question is just talking about four little numbers, like 1, 3, 5, and 7.

19 When they invent a strange-looking symbol and put it in a math question, focus on their definition of that symbol. Then work through it, step-by-step.

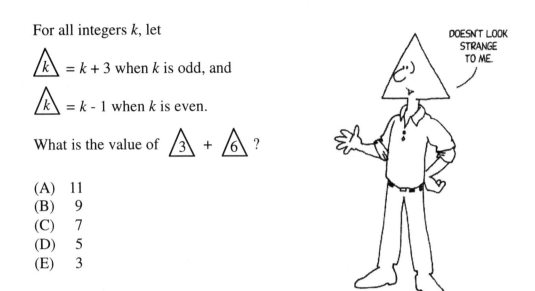

For all integers *k*, let

$\triangle k$ = *k* + 3 when *k* is odd, and

$\triangle k$ = *k* - 1 when *k* is even.

What is the value of $\triangle 3$ + $\triangle 6$?

(A) 11
(B) 9
(C) 7
(D) 5
(E) 3

Don't be scared! If ever there was an extremely easy math question dressed up to look like something, this is it. First they put letters inside triangles, then they ask you to add numbers inside triangles! What kind of crazy math is this?

It's SAT math. And it isn't hard at all. It just looks like something you may have never seen.

They're creating a symbol. And they're defining it for you by giving you two rules:
• When you put an odd integer inside the triangle, the value of that expression is *the odd integer plus 3*.
• When you put an even integer inside the triangle, the value of that expression is *the even integer minus 1*.

Got it? Okay, when you put a 3 inside the triangle, which rule do you use? Right, the first one, because 3 is odd. So:

$\triangle 3$ = 3 + 3, or 6.

And:

$\triangle 6$ = 6 - 1, or 5.

So the answer is 6 + 5, or 11 (**A**).

Did you notice?	The answers are all in the form of ordinary numbers, and you get the answer in just a few steps. So on an actual SAT, this question would have been closer to the beginning of the section, maybe number 8 or 9. If the answers had been in the form of the symbol (numbers inside triangles), the question would require more steps, and would have appeared later in the section. For example, see question 24.

20

Don't just solve for the variable -- answer the question. (Part II)

How many different integers m will make the following statement true?

$4 < 3m < 8$

(A) One
(B) Two
(C) Three
(D) Four
(E) Five

Be careful! This is a classic example of the kind of question too many people answer quickly, confidently, and often, incorrectly. What makes such an easy question so easy to mess up? Look:

We're give a question, a statement, and a set of answers.

The question is: "How many different integers make the statement true?"

Not "Which ones?"

But rather "How many?"

The statement is: $4 < 3m < 8$

For this part, your brain naturally shifts into solve-for-the-variable mode. It wants to know what m equals. This is the kind of question you're accustomed to answering. So you tend to slide more easily into this mode. Which is what the Emperors of Evil at SAT headquarters are hoping for.

See, there's only one possible solution for the statement. The product of 3 and m must be greater than 4 and less than 8. Since 3 x 1 is less than 4, m cannot equal 1. And because 3 x 3 is greater than 8, m cannot equal 3, or anything larger than 3. The only solution for m that makes the statement true is $m = 2$, because 3 x 2 is greater than 4 and less than 8.

Your brain probably did all that thinking in about one second. But in that one second, it may have also decided that the answer to the *question* is 2 and begun a dangerous leaning toward choice B.

Not good. Again, the question is "How many different integers make the statement true?" The answer to that question is "one." There is only **one** integer m that makes the statement true. And that **one** integer happens to be 2. But the answer to the question is **one**. Choice **A**.

Did I drag that out too much? Did you know the answer was A way back at the beginning? Sorry, I get a little over-protective sometimes.

| **Did you notice?** | Those *less than* and *greater than* signs make you just a little edgy, don't they? Even though we've seen them a lot, we still don't completely trust them. (You have to keep an eye on them, unlike our good friend the *equal* sign, which seems, I don't know, more honest.) And once again, we have numbers spelled out as words in the answers. But you can disarm that trap just by noticing it. |

To find the area of an irregularly-shaped region, first find the area of the whole, then subtract the unwanted sections. (Part I)

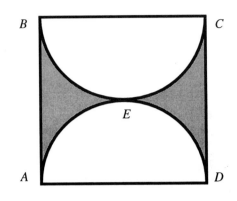

In rectangle *ABCD* above, semicircles *AED* and *BEC* meet at point *E*. If the radius of each half-circle is 1, what is the area of the shaded region?

(A) 1 - π/2   (B) 2 - π   (C) 4 - π/2   (D) 4 - π   (E) 2

Almost all of these "area of the shaded region" problems require the same strategy. First find the area of the whole thing (in this case the rectangle). Then find the area of the parts you're not interested in (the semicircles). Finally, subtract the area of the semicircles from the area of the rectangle. What you'll have left is the area of that strange-looking shaded region.

In order to solve this problem, we'll need to know how to find the area of a rectangle ($l \times w$) and the area of a circle (πr^2).

Do you see that the length of the rectangle is also the diameter of the semicircle? Since the radius of the semicircle is 1, the diameter is 2. Therefore, the length of the rectangle is 2. Going down, we see that the radius of the top semicircle plus the radius of the bottom semicircle equals the width of the rectangle. So the width of the rectangle is also 2, and we now know it's a 2 x 2 square. The area of this square is 2 times 2, or 4.

If we put the two semicircles together, we have one whole circle, with radius 1. Plugging this information into the formula, we see that the area of this whole circle is $\pi (1^2)$, or just π.

When we subtract the area of the two semicircles (π) from the area of the rectangle (4), we find the area of the shaded region to be 4 - π. Answer **D**.

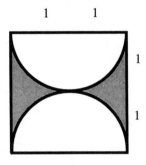

Length of rectangle = 2 x radius of circle
Width of rectangle = 2 x radius of circle

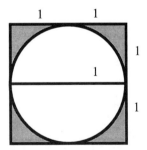

Area of rectangle = 2 x 2 = 4
Area of circle = $\pi r^2 = \pi (1) = \pi$
Area of shaded region = 4 - π

Did you notice?

If we had tried to find the areas of the two semicircles individually, we would've been dealing with multiplying πr^2 by 1/2, then multiplying the result by 2. Assuming no errors, we would have ended up in the same place, but after more time and work. Whenever possible, make it easier on yourself.

22

Right triangles? Lengths of sides? Pythagorean Theorem!
Figure not drawn to scale? They're not kidding!

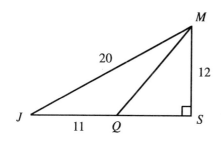

Note: Figure not drawn to scale.

In triangle *JMS* above, what is the length of *QM*?

(A) 16   (B) 13   (C) 12   (D) 10   (E) 5

Once again, we're dealing with a right triangle, and the question is asking about the lengths of the sides. Automatically, we should pull the good old Pythagorean Theorem out of the mental toolbox. In all right triangles, $a^2 + b^2 = c^2$. Interest rates may climb, the market may fall, your team may fold, but you can always count on the Pythagorean Theorem. Without a doubt, $a^2 + b^2$ will always equal c^2.

In the above problem, notice there are two right triangles, *JMS* and *QMS*, and that one is inside the other. Looking at the larger triangle, we know the lengths of *JM* and *MS*. Once we've found *JS*, we can then subtract *JQ* from it to find *QS*. And once we know *QS*, we will have two sides of the smaller right triangle (*QMS*), and you already know how to find the third side.

Let's apply the Pythagorean Theorem first to △*JMS*.

$(12)^2 + (JS)^2 = (20)^2$

$144 + (JS)^2 = 400$

$(JS)^2 = 256$

$JS = 16$
(WATCH OUT! The answer is not A!)

Now let's move to △*QMS*. ✒

Since *JS* = 16, and *JQ* = 11, then *QS* = 5

(WATCH OUT AGAIN! The answer is not E!)

$(QS)^2 + (MS)^2 = (QM)^2$

$(5)^2 + (12)^2 = (QM)^2$

$25 + 144 = (QM)^2$

$169 = (QM)^2$

$13 = QM$ (RELAX! The answer is **B**!)

Did you notice?

"Figure not drawn to scale." This means that just because *JQ* looks like it's equal to *QS* doesn't mean it is. In fact, you can be pretty sure the lengths are different. This "figure not drawn to scale" thing is done partly to prevent you from simply measuring the lengths and then picking a good answer. It's also done partly to mess you up. So don't resort to measuring. And don't let them mess you up!

23

To find the slope of a line: pick any two points on the line, subtract the y's from each other, subtract the x's from each other, and divide the differences (y on top).

What is the slope of the line that passes through the origin and the point (4,2)?

(A) 0   (B) 1/2   (C) 1   (D) 2   (E) 4 1/2

Is the concept of slope confusing for you? Let's clear it up.

Think of a ski slope. As you travel down the mountain, you're moving in two directions (the first time I skied, I moved in eleven directions, but I don't feel like talking about that now). Beginning at the top of the hill, your position can be defined as a point on a graph. As you ski downhill, your position continues to change all the way to the bottom. At any given point on the hill, you have a new set of coordinates in relation to your starting position at the top. Using those two points, you could determine the *slope* (or steepness) of the line you've traveled. Depending on the slope of your path, you will move a certain number of feet vertically for every foot you move horizontally.

In the example above, let's say the top of the hill is the point (4,2). On your way down the mountain, you pass through the point (0,0). In order to find the slope, we're going to subtract the two *y* values from each other. Because it's easier, let's do 2 minus 0. That equals 2. Since we subtracted 2 minus 0 for the *y*, we have to follow the same order and use 4 minus 0 for the *x*.

So we say "the change in *y*" is 2 (we went from 2 to 0), and "the change in *x*" is 4 (we went from 4 to 0). When we divide the change in *y* by the change in *x*, we get:

$$\text{Slope} = \frac{2 - 0}{4 - 0}$$

$$= \frac{1}{2}$$

Answer **B**.

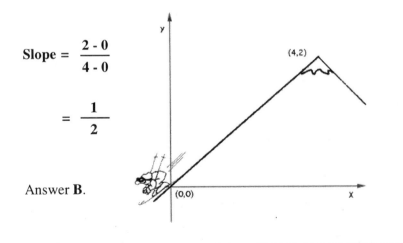

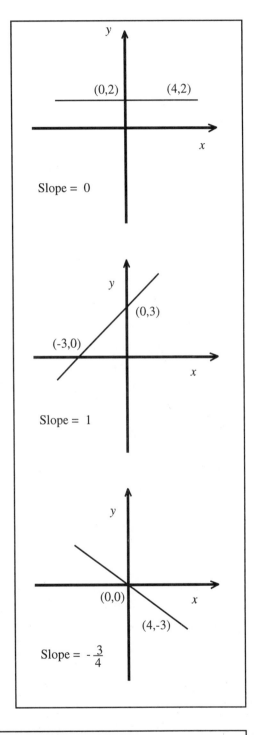

Slope = 0

Slope = 1

Slope = $-\frac{3}{4}$

Did you notice?

The slope can be a whole number or fraction, positive or negative. If the line is parallel to the *x*-axis (horizontal), the slope is 0. But if it's parallel to the *y*-axis (vertical), we say that it has no slope. This is because wherever you go on that line, *x* has the same value. Therefore, the "change in *x*" is zero, and because we can't have zero in the denominator of a fraction, the slope cannot be expressed.

When the answers contain the weird symbol, continue for another step or two. It's just a little more of the same thing.

For all integers z, let Ⓩ be defined by Ⓩ $= z^2 + 3$. Which of the following is equal to ④ $-$ ③ ?

(A) ①
(B) ②
(C) ③
(D) ⑤
(E) ⑦

This is another one of those weird symbol problems that has contributed greatly to the SAT's reputation as a beloved American institution. It is similar to the one we met up with on page 19, but with an important added attraction. The above question requires an extra step, because the answers themselves have to be translated. That's why this one would likely appear later in the section -- because it's "harder" (not really, just more steps). Otherwise it's the same problem.

According to the definition given, we find the value of the number-inside-the-circle thing by adding 3 to the square of the number. Therefore:

④ $= 4^2 + 3 = 19$

and

③ $= 3^2 + 3 = 12$

Then: ④ $-$ ③ $= 19 - 12 = 7$

Because the answers to this question are expressed in terms of the weird symbol, we now have to translate the answer we found, 7, into one of the answers given. In other words, which of the answers is equal to 7? With a little trial-and-error, we quickly discover that **B** is the correct answer, because:

② $= 2^2 + 3 = 7$

Did you notice?	This question tends to scare many people, because at first glance, even the answers look strange. But as you've seen, solving it is simply a matter of doing a bunch of easy little steps. Always be careful not to choose an answer like E just because it's a number you happened to come up with on your way to the *real* right answer. The incorrect choices are there for a reason. Don't get trapped!

25 If a square is formed by connecting the centers of circles, there will be a strong relationship between the radii of those circles and the length of that square.

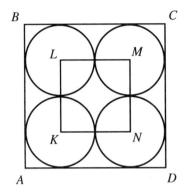

The area of square *ABCD* above is 16, and points *K*, *L*, *M*, and *N* are the centers of their respective circles. What is the area of square *KLMN*?

(A) 2
(B) 4
(C) 6
(D) 8
(E) 10

Work backwards! If they want the area of a square, the last step before you arrive at the answer will be to determine the length of the square.

Obviously, the circles have something to do with the length of square *KLMN*. We know that all four circles are the same size, because that's the only way they could fit so nicely into the larger square. So the radii of the circles are equal. That means if we knew the radius of one circle, we could double it and that would be the length of square *KLMN*.

Okay, we're getting warmer. Let's now use what they gave us. The area of square *ABCD* is 16. Therefore, the length of the larger square is 4. See that? Now don't get lost. The length of square *ABCD* is also equal to the diameters of two of the circles -- or four of the radii. Since the length of square *ABCD* is 4, and it equals four radii, then one radius must be 1.

The length of square *KLMN* is 2 x 1, or 2. But watch out -- that's not the answer! The area of square *KLMN* is 2 x 2, or 4 (answer **B**).

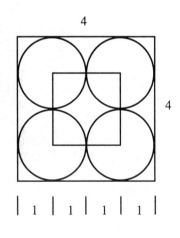

Did you notice? All of the answers given are possible, but because 4 is the only perfect square among them, it's a good bet that will be the correct answer. (In a similar way, the volume of a cube must be the cube of a number. Very often the answer will be 8, 27, 64, or 125.) Back to the question above: if they had asked for the area of that curvy diamond shape inside square *KLMN*, would you know how to find it?

Stay focused on what they're asking. They're hoping
you'll give the right answer to the wrong question.

A certain rectangle is divided in half so that two
squares are formed. If each square has a perimeter
of 36, what is the perimeter of the rectangle?

(A) 162   (B) 72   (C) 63   (D) 54   (E) 36

First thing: draw a picture!
   It should look something like this:

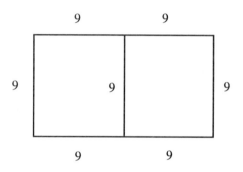

What are they asking for? The perimeter of the original rectangle.
   What are they hoping you'll do? Add the perimeter of the square on the left to the perimeter of the
square on the right, come up with 72, and select B as your answer. Or, forget the difference between
perimeter and area, multiply 9 times 18, come up with 162, and select A.

First thing: the only way you can divide a rectangle into two squares is if one side is twice as long as the
other, as shown. Notice they don't give you a diagram for this question, because they're also hoping you'll
have trouble drawing the picture.

Second: if the perimeter of each square is 36, then the side of each square must be 9.

Third: the perimeter of the rectangle is what they're asking for. But the wording of the question has your
mind focused on the perimeters of the squares. It's like hypnosis. Stay awake! The perimeter of the
rectangle is 9 + 18 + 9 + 18, or 54. Answer **D**. That middle line doesn't count, not twice, not once, not at all.

Did you notice?	Every answer given is there for a reason. We've already explained A and B. Answer C is there because some people will simply add up all the sides and that center dividing line, and come up with 63. Answer E is the perimeter of one of the squares, which of course is not what they're asking for.

To find the area of an irregularly-shaped region, first find the area of the whole, then subtract the unwanted sections. (Part II)

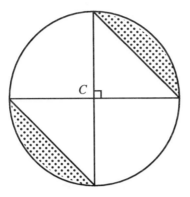

The circumference of the circle with center at *C*, shown above, is 8π. What is the area of the shaded region?

(A) 8π (B) 16π (C) $8\pi - 16$ (D) $4\pi - 8$ (E) $4\pi - 16$

They want us to find the area of those two little slices of the circle. Do you see that if we found the area of the entire circle, we could figure out the area of each quarter-circle (the triangle plus the slice)? Then, we could figure out the area of the triangle and subtract that to find the area of just the slice. When we double the result, we have the area of both slices. I know it seems confusing, but it isn't. (Okay, maybe it is, but you can still do it!)

Because we'll need to find the area of the circle, we must first determine the radius of the circle. We're told the circumference is 8π. The formula for circumference is $C = 2\pi r$. So if $2\pi r = 8\pi$, then the radius *r* must equal 4. Now just follow these five steps:

1. Area of Circle

The area of any circle is given by the formula πr^2 Since $r = 4$, the area of the circle is $4^2\pi$, or 16π.

2. Area of Quarter-Circle

If the area of the circle is 16π, then the area of the quarter-circle is 4π. (We know it's a quarter-circle by the right angle.)

3. Area of Triangle

The radius, 4, is also the length of the two legs of the right triangle. The area of triangle is 1/2 base times height. In this case, it's 1/2 (4 x 4), or 8.

4. Quarter-Circle minus Triangle (One Slice)

$4\pi - 8$

5. Area of Both Slices

$8\pi - 16$. Answer **C**.

Did you notice?

That was a lot of work for one point. But if you followed step-by-step, I think you saw that this problem is like so many others: just a matter of doing a series of easy tasks in a logical order. And just like the others, this one contains incorrect answers that match the results you get along the way. As you do each step, remind yourself, quickly, where you are and where you're trying to go.

Take the route that's best for you. If your mind handles numbers easily, use arithmetic. If it thinks better in symbols, use algebra.

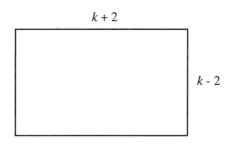

$k + 2$

$k - 2$

If the area of the above rectangle is 45, what is the value of k?

(A) 5
(B) 7
(C) 9
(D) 13
(E) 17

What piece of knowledge do you need to solve this problem? You need to pull from your mental toolbox that the area of a rectangle is obtained by multiplying the length times the width.

Here's the arithmetic approach:

The length and width, when multiplied together, must equal 45. So the length and width must be factors of 45. The factors of 45 are 1, 3, 5, 9, 15, and 45. So there are three pairs of factors: 1 x 45, 3 x 15, and 5 x 9. Since, according to the diagram, one factor is obtained by adding 2 to k and the other by subtracting 2, the factors must be pretty close together. We could plug in each pair, but if we guessed at 5 and 9, we'd be right, because 7 plus 2 is 9 and 7 minus 2 is 5. Therefore, k must equal 7.

(Answer: **B**)

Here's the algebra approach:

The length and width, when multiplied together, must equal 45. So in algebraic terms, $(k + 2)(k - 2) = 45$. Then

$$k^2 + 2k - 2k - 4 = 45$$

$$k^2 - 4 = 45$$

$$k^2 = 49$$

$$k = 7$$

Did you notice? Although neither 5 nor 9 is the correct answer, both are answer choices. They are placed there because the testmakers know a certain percentage of people will be overjoyed that a number they came up with in their work actually appears in the answers, and they will select the answer for no other reason. Don't do that.

29 The three angles of any triangle -- even a triangle within a triangle -- add up to 180 degrees.

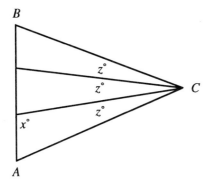

If, in isosceles △*ABC* above, angle *BAC* = 63, then *x* equals

(A) 54   (B) 63   (C) 96   (D) 99   (E) 126

The key here is to understand what an isosceles triangle is, to stay focused on the fact that the angles of every triangle total 180 degrees, and to label the diagram as you go along. Remember also that all of those *z*'s are equal to each other. (That's true in this and in every math problem.)

In an isosceles triangle, two of the sides are equal, and the two angles opposite those sides are equal. We can see from the diagram that *AC* = *BC*. Therefore, ∠*ABC* = ∠*BAC*. We're told that ∠*BAC* equals 63 degrees. So ∠ABC also equals 63.

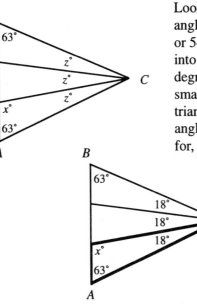

Looking at △ABC, we see that the two known angles add up to 126. So ∠*ACB* must be 180 - 126, or 54 degrees. And because that angle is divided into three equal parts, one of the parts, *z*, equals 18 degrees. Now look at just the lowest of the three small triangles. We know two of the angles in that triangle: 63 and 18. They add up to 81. So the third angle, which happens to be the one we're looking for, must be 180 - 81, or 99 degrees (Answer **D**).

| **Did you notice?** | There is no warning: *Figure not drawn to scale*. That means you can rely on your own eyes to tell you the angle in question is more than 90 degrees (can you see that it's greater than a right angle?) Which means answers A and B can be eliminated immediately. Also, even though △*ABC* contains 180 degrees, the three smaller triangles within it also contain 180 degrees each. |

30 Read the question quickly, then analyze the chart or graph so that you understand what it's saying. Then read the question again. (Part I)

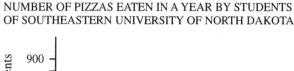

NUMBER OF PIZZAS EATEN IN A YEAR BY STUDENTS OF SOUTHEASTERN UNIVERSITY OF NORTH DAKOTA

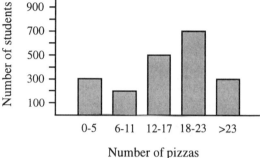

According to the graph above, which of the following is closest to the percentage of students from Southeastern University of North Dakota who eat at least 18 pizzas a year?

(A) 20%   (B) 25%   (C) 30%   (D) 40%   (E) 50%

I want you to take several minutes to study the bar graph above. It isn't that it's so complex or difficult to understand. It's just that I spent a really long time creating that bar graph and I want to make sure you appreciate it.

You've read the question. You know it concerns something about the number of pizzas eaten by some group of students at a midwestern university. What are they asking? They want the percentage of students who eat at least 18 pizzas. Remember: "at least 18" means "18 or more." Which of the bars in the graph represent 18 or more? The last two, the ones labeled "18-23" and ">23."

The bar labeled "18-23" represents 700 students. The bar labeled ">23" represents 300 students. So 1,000 students at this university eat at least 18 pizzas a year. But the question (and the answers) have to do with percentages. How can we find out what percentage of the entire student body 1,000 students is? We have to know the total number of students there are in the university.

If we examine the bars in the graph, we soon see that every student must fall into one of the five categories. Adding the numbers represented by the five bars, we come up with a total of 2,000 students. So 1,000 is one-half of the students, or 50% (**E**).

I guess the real question is, what's wrong with the other half?

| **Did you notice?** | Like so much of math, this graph is about relationships: how big is one group compared to another? Once you understand the graph, the question is simple. Out of 2,000 students, 300 eat from 0 to 5 pizzas in a year, 200 eat from 6 to 11, 500 eat from 12 to 17, 700 eat between 18 and 23, and another 300 eat more than 23. How many eat 18 or more pizzas? 1,000. And 1,000 is 50% of 2,000. |

31

A percent is just a fraction that has been changed to an equivalent fraction with 100 in the denominator.

THE ONLY THING
BETTER THAN
MY MATH
IS
MY COOKIES!

Ida has a mixture consisting of 12 pounds of flour and 4 pounds of sugar. What is the minimum number of pounds of sugar she must add to end up with a mixture that is 40 percent sugar?

(A) Two   (B) Three   (C) Four   (D) Five   (E) Six

Don't be intimidated by percents. Remember, percents are just like fractions, ratios, and probability: a part of the whole being compared to the whole. The trick with this particular question — and many variations of it — is remembering what the whole really is.

The original mixture had 4 pounds of sugar and 12 pounds of flour, for a total of 16 pounds. So the ratio of sugar to the whole mixture was 4:16, or 4/16, or 1/4, or .25, or 25%. They all mean the same thing. Forty percent just means that the amount of sugar over the whole mixture, as a fraction, will equal 40/100.

So let's call x the number of pounds of sugar we must add to the mixture. The new amount of sugar, then, will be $4 + x$. And the new total for the whole mixture will be $16 + x$ pounds. (Remember, the original mixture was 16 pounds and now we're adding x pounds of sugar to that.)

Therefore,

$$\frac{4 + x}{16 + x} = \frac{40}{100}$$

When we cross multiply, we get

$100 (4 + x) = 40 (16 + x)$

$400 + 100x = 640 + 40x$

$100x = 240 + 40x$

$60x = 240$

$x = 4$ (Answer C)

An alternative approach:

Because the answers are small, you can plug each one into the left side of the equation and check to see which one equals 40 percent.

What you'll find is that:

$$\frac{4 + 4}{16 + 4} = \frac{40}{100}$$

$$\frac{8}{20} = \frac{40}{100}$$

Plugging 4 in for x makes the equation work, because 8/20 equals 40/100.

Did you notice?

The problem gives numerals but the answers are written in words. Have you ever gotten the right answer on a question, and then picked the wrong answer for some mysterious reason? This tactic just increases the possibility that you'll do that. Also, in this type of word problem, be sure you add the additional amount to the numerator *and* the denominator. Both are increasing.

Think of ratios as shares. A 3:2 ratio means one part has three shares for every two shares the other part has.

The measures of the four angles of a quadrilateral are in the ratio 3:2:2:1. What is the measure of the <u>largest</u> angle?

(A) 30   (B) 45   (C) 60   (D) 90   (E) 135

Do you know what a quadrilateral is? It's a four-sided polygon. And since it has four sides, it also has four angles. This problem is telling us that those angles of one particular quadrilateral have degree measures in the ratio of 3:2:2:1. Now what does that mean?

It means that the largest angle is three times the measure of the smallest one, and that two of the angles are equal to each other and twice the size of the smallest. Like so much of math, ratios are about relationships -- in this case, the relationships among the sizes of the various angles in this polygon.

What is this ratio really saying? That if you divided all the degree measures in the angles of a quadrilateral into equal pieces, the largest angle would have three of those pieces, the smallest one piece, and the two middle angles two pieces each. But how do we find out how big the pieces are?

We know a quadrilateral has four angles, and that the angles add up to 360 degrees. (Need proof? Take any regular four-sided polygon and divide it into two triangles. Remember that every triangle has 180 degrees. So <u>two</u> triangles must have 360 degrees.

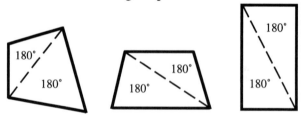

Now here's a good trick. See those numbers in the ratio (3:2:2:1)? Add them up (3+2+2+1). They add up to 8. So we're going to take the whole (360 degrees) and divide it into 8 pieces. 360 divided by 8 is 45. That means that each piece is 45 degrees. The smallest angle has one of those pieces (1 x 45), or 45 degrees. Each of the two middle angles has two pieces (2 x 45), or 90 degrees. And the largest angle has three pieces (3 x 45), or 135 degrees. Answer **E**.

Don't trust this technique? Check it: 45 + 90 + 90 + 135 should equal 360.

Try another one. "The three angles of a triangle are in the ratio 1:2:3. What's the largest angle?" The whole is 180 degrees. Now add 1 + 2 + 3. That equals 6. Then divide 6 into 180 to get 30. So the smallest angle is 30, the middle is 2 times 30, or 60, and the largest is 3 times 30, or 90. And sure enough, the three angles of the triangle (30 + 60 + 90) add up to 180 degrees. Works every time.

Did you notice?	The result you get when you divide 360 by 8 is 45. Many people (not you) stop thinking at this point and choose B as their answer. Remember to answer the question, and this question is asking for the largest angle. The largest angle in this problem has three pieces, or shares, of 45 degrees each. Don't do all that work, then pick the wrong answer!

33 Break word problems into parts. Then work, step-by-step, from what you know toward what you're trying to find out.

If 1 can of brand X dog food feeds 4 puppies or 2 adult dogs, 8 cans of brand X dog food will feed 24 puppies <u>and</u> how many adult dogs?

(A) Two
(B) Four
(C) Eight
(D) Twelve
(E) Sixteen

This is the kind of word problem that could cause you to become an animal hater. It's also the kind of question you can read over and over and over, and if you don't step back from it, you might never get any closer to the answer. (My immediate response is to get rid of the dogs and go buy some goldfish. However, in the interest of helping you out with the SAT, I'm going to address the question as it has been presented.)

You and I are out shopping for dog food. We have a coupon for brand X, so that's the kind we're going to buy. (I learned this from my mother. The stuff inside the cans could be blue, but if you have the coupon, use it.) We have exactly enough money to buy 8 cans of dog food. With this feast, we have to feed 24 puppies and as many of the adult dogs as we can. Then, while I wait outside, you will calmly explain to the other adult dogs that as soon as we get more money, they can have some food.

We were told that 1 can will feed 4 puppies. So we'll need 6 cans to feed all 24 puppies. Remember, we started with 8 cans, and now we've used 6 of them up on the little guys. We have 2 cans left. Since each can will feed 2 adult dogs, and we have 2 cans, we can feed 4 adult dogs. The answer is **B**.

That's it. As usual, what we have here is a simple math problem dressed up in the confusing disguise of SAT language.

Did you notice?	The numbers appear as numerals in the question, but as words in the answers. So another part of your brain has to take over to find the answer you've already found. Why would they throw an X (brand X) into a math problem when it doesn't need to be there? And what about the words *or* and *and* in the question? Doesn't this all create more confusion? (Only if you let it.)

The "fence post" question usually tries to lure you to an answer that is 1 less than the correct choice.

How many vertical posts, spaced 5 feet apart, would be needed to build a free-standing fence 40 feet long?

(A) Five
(B) Six
(C) Seven
(D) Eight
(E) Nine

Think first, then do the math! This would appear somewhere in the middle of a math section. It wouldn't be question number 2, nor would it be question number 22. It isn't difficult, but it's tricky -- because it appears to be so simple.

It's obvious they want you to divide 5 into 40, come up with 8, and choose answer D. But an alarm should go off in your head: This is question number 15. There should be more work involved. In other words, that was a little too easy.

Think about it. If all they wanted you to do was divide 40 by 5, why would they go through all the trouble of writing a word problem, then just give you the only two numbers you'll need to answer the question? Unless you're brand new to the SAT and you just jumped right to this page, you know better by now. The SAT people may be extremely generous to each other and to their families, but they're not about to drop a gift like that into your lap. Not while you're awake, anyway.

So where does that leave us? With the need for a quick picture.

The fence should look something like this (although you shouldn't spend the time making yours quite so stunning):

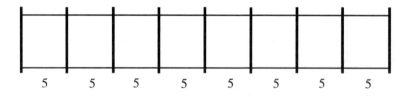

5   5   5   5   5   5   5   5

There's probably no need for a detailed explanation. As you can see, the fence contains eight 5-foot sections, for a total of 40 feet. The key to this question is the fact that there's a post at the beginning and end of each section. So you need two posts to complete one section, three posts for two sections, and so on. In the end, you need nine posts to build 8 sections. Answer **E**.

| **Did you notice?** | Your mind immediately wants to calculate the number of 5-foot sections, rather than the numbers of posts, so you can easily arrive at an answer that is off by one. Also, watch out for other forms of the fence post question (see page 76). All involve intervals -- sections of fence, periods of time. Just remember that all intervals, including the first and last ones, have a beginning and an end. |

35

Substitute to get rid of the variables you don't want.

If $a - b = n$ and $a + b = p$, then in terms of n and p, which of the following is equal to ab?

(A) $p - n$ (B) $\dfrac{p - n}{2}$ (C) np (D) $\dfrac{p^2 - n^2}{2}$ (E) $\dfrac{p^2 - n^2}{4}$

The first step in solving this problem is determining what they're asking. What do they mean, "In terms of n and p, which of the following is equal to ab?"

They want you to re-arrange the two given equations so that a equals something and b equals something, and when you multiply those two somethings together (a times b), you get an answer that only has the variables n and p in it. That's what they mean by ab "in terms of" n and p.

To accomplish this, we'll have to move a few things around, then substitute.

If:	$a - b$	$= n,$	If:	$a - b$	$= n,$
then	a	$= n + b$	then	b	$= a - n$

Let's plug this new information into the other equation:

$$a + b = p$$
$$(n + b) + b = p$$
$$n + 2b = p$$
$$2b = p - n$$
$$b = \frac{p - n}{2}$$

Let's plug this into the other equation:

$$a + b = p$$
$$a + (a - n) = p$$
$$2a - n = p$$
$$2a = p + n$$
$$a = \frac{p + n}{2}$$

Now we have a and b in terms of n and p. Let's multiply.

$$ab = \frac{p + n}{2} \cdot \frac{p - n}{2} = \frac{p^2 + np - np - n^2}{2 \cdot 2} = \frac{p^2 - n^2}{4} \quad \text{(Answer E.)}$$

# Did you notice?	If you were to forget to multiply the denominators (2 • 2), you would have done a lot of work, only to come up with the incorrect answer D. Many people get almost to the end, then multiply $(p + n)$ and $(p - n)$ in their heads, and smartly jump right to $p^2 - n^2$. But they blow the whole thing by failing to multiply the denominators. Be careful!

In word problems involving rate, time, and distance, begin with the formula.

If Joe and Noreen drive to New York at an average speed of 60 miles per hour, they will arrive 3 hours early. If they average 40 miles per hour, they will arrive 3 hours late. What is the length, in miles, of the round-trip?

(A) 12   (B) 15   (C) 18   (D) 720   (E) 1,440

This is a deliciously wicked question, designed to cloud your brain and make you wish you had been born into a royal family in Scandinavia. But don't despair. It turns out to be a fairly simple question to answer, if you promise to be careful: the traps are numerous.

Let's begin with the formula. Rate x Time = Distance

Now let's shorten it to $R \times T = D$. The problem talks about arriving 3 hours early and 3 hours late, so there must be a number of hours that would be considered "on time." Let's continue to call that T. According to the question, plugging in a rate of 60 would decrease T by 3 hours. Plugging in a rate of 40 would increase the time by 3 hours. So,

60 (T - 3) = D and 40 (T + 3) = D

Since the distance remains the same throughout the problem (unless they move New York), D is always the same. Therefore,

60 (T - 3) = 40 (T + 3)

60T - 180 = 40T + 120

20T = 300

T = 15

Now be careful. The question is asking for the distance, not the time. So we need to plug 15 back into one of the equations. For example, if 40 (T + 3) = D, and T = 15, then the distance is 40 x 18, or 720 miles.

Be careful again! They want the round-trip distance. If it's 720 miles to New York, the round-trip is 1,440 miles. The answer is **E**.

Did you notice?	Even if you were an absolute genius at setting up the equation, there are still at least four places where you could mess up -- not even counting the possible arithmetic errors. Remember to answer the question. They're not asking you to find T, or T + 3, or T - 3, or even D. They're asking for the round-trip distance. But as always, those other values are well-represented in the answer choices.

Don't let a few mysterious letters cause you to forget how to add.

```
  1 P 5
    Q 4
+   R 2
─────────
  3 2 1
```

In the addition problem above, the digit *R* could be equal to which of the following?

I.   1
II. 6
III. 9

(A) II only   (B) III only   (C) I and III only
(D) II and III only   (E) I, II, and III

ROMAN NUMBERS.
ARABIC NUMERALS.
ROMAN LETTERS.
IT'S ALL GREEK TO ME!

This is often a strange, impossible-looking question for the newcomer, who is tempted to choose E and move on. However, let me assure you that to do so would be a mistake, because E is not the right answer.

Let us begin at the beginning. Those letters you see (*P*, *Q*, and *R*) in that addition problem represent whole numbers. There are only ten possibilities for each of them, namely the digits 0 through 9. (It might help to see this as a 2nd-grade arithmetic question dressed in SAT clothing.)

Look at the right-hand column of numbers: 5 + 4 + 2. They add up to 11. So the "1" under that column is really 11, and a "1" has been carried over to the ten's column. That means that *P* + *Q* + *R* + 1 equals some number that ends in a 2 (see that?)

```
   2 1
  1 P 5
    Q 4
+   R 2
─────────
  3 2 1
```

If we look to the far left column of numbers, we see that a "2" has been carried over so that when it's added to the "1" we get a "3" in the sum. That means the middle column must total 22. Since we carried a 1, *P*, *Q* and *R* must add up to 21. Now the most P and Q can total is 9 + 9, or 18. So in order for the three numbers to add up to 21, *R* must be at least 3.

Obviously, *R* cannot equal 1, so statement I does not work. The answer is **D**.

Did you notice?	We didn't bother to check if 6 and 9 really worked. Once we discover the truth about this question -- that the highest sum for any two digits is 18 -- we realize the third digit must be at least 3. Therefore, *R* can be any of the digits 3, 4, 5, 6, 7, 8, or 9. Remember: we aren't solving to see what the letters do equal; we're solving to see what they could equal. In this problem, they can't equal 1 or 2.

Sometimes you can identify the correct answer
without doing any calculations.

If $x = 25$, then $x^2 - \sqrt{x}\ =$

(A) 75
(B) 195
(C) 528
(D) 620
(E) 743

Here's a great example of a problem that is easy enough to do, either in your head, on paper, or with the calculator. But it's also a chance to learn a little mathematical technique — another tool for your toolbox — that can come in very handy on more complex problems.

When you square 25, you get 625. In fact, when you square any number that ends in a 5, you get an answer that also ends in a 5. If you want the whole truth, when you raise a number ending in 5 to any power, the result will end in a 5. Every time. Since the square root of 25 is 5, the question is asking you to subtract 5 from a number that ends in a 5. The result will always be a number that ends in a 0, because that final 5 minus 5 will always give you a zero at the end of your answer. See what I mean? Look at the answer choices and you will notice a most wonderful thing: there's only one answer that ends in a zero. So that has to be it: **D!**

Is this really just a trick? No, it's math. And it's been built into the question, because otherwise there would be other answers that end in zero.

Okay, you're thinking, it's math, but is it a helpful tool? Well, what if the question had looked like this:

If $x = 25$, then $x^5 - \sqrt{x}\ =$

(A) 125
(B) 672,812
(C) 8,765,323
(D) 9,765,620
(E) 9,765,625

Now what are you going to do? Even with a calculator, this is going to take a little time. And with those last three choices looking so much alike, you're going to end up doing it twice just to make sure. And then, if you end up with two different answers, you have to start all over (don't you hate that?) Try both methods.

| # Did you notice? | And there are others. For example, any number ending in a 6, when raised to any power, will always yield a result also ending in a 6. (I don't know what you get when you raise 11,427,876 to the 9th power, but I know it ends in a 6. And I find that comforting, somehow.) |

When one geometric shape is inscribed in the other, their common feature is often the connection between what you're given and what you need to find.

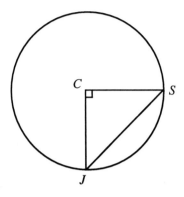

The area of the circle with center C, shown above, is 60π. What is the area of triangle JCS?

(A) 4   (B) 7   (C) 30   (D) 55   (E) 94

Let's take a step back from this question and look at what we're given and what they're asking.
>   They want the area of the triangle.
>   In order to find the area of the triangle, we need to know the lengths of the base and height.
>   The base and height of the triangle correspond to the radius of the circle.
>   If we could determine the radius, we would know the base and height of the triangle.
>   We could determine the radius if we knew the area of the circle.
>   They've given us the area of the circle!

Now work backwards. The area of the circle is 60π. Using the formula for area of a circle, we get the radius:

$$A = \pi r^2 \quad = 60\pi$$
$$r^2 \quad = 60$$
$$r \quad = \sqrt{4} \cdot \sqrt{15} \ = \ 2\sqrt{15}$$

Because the legs of the right triangle are formed by the radius of the circle, we can now plug $2\sqrt{15}$ right into the formula for area of a triangle:

$$A \ = 1/2\ \text{base} \cdot \text{height}$$
$$= 1/2\ (2\sqrt{15} \cdot 2\sqrt{15}\ )$$
$$= 30 \ \text{(Answer C)}$$

Did you notice? You can only use the legs of the triangle as base and height if it's a right triangle. And they don't tell you it's a right triangle -- they just indicate it in the diagram. Also, the mistake most people make is forgetting to multiply by *one-half*. When you neglect to use that one-half, you get an answer that's twice what it should be. Whether or not that wrong answer is one of the choices, you have a problem.

40

If the question asks for the circumference of a circle, the next-to-the-last step must be to find the radius of the circle.

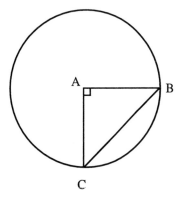

In the circle with center A, above, the length of BC is 6.
What is the circumference of the circle?

(A) 6π (B) 12π (C) $2\sqrt{3}\,\pi$ (D) $3\sqrt{2}\,\pi$ (E) $6\sqrt{2}\,\pi$

Hey, isn't this the same as question 39? Same picture, different question. Here they've given us the length of the hypotenuse of that right triangle, and they want the circumference of the circle. So we need to find the radius. Notice that the radius of the circle is again the length of the two legs of right triangle ABC. If we can find the length of one of those legs (which we can), we'll be there. Since we need to find the lengths of the sides of a right triangle, let's pull out the Pythagorean Theorem: $a^2 + b^2 = c^2$. Or, in this case:

$$(AB)^2 + (AC)^2 = (BC)^2$$

Then, because BC equals 6,

$$(AB)^2 + (AC)^2 = 36$$

AB and *AC* are radii of the same circle,
so they must be equal to each other.

$$r^2 + r^2 = 36$$

$$2r^2 = 36$$

$$r^2 = 18$$

$$r = \sqrt{9} \cdot \sqrt{2} = 3\sqrt{2}$$

Don't forget this step!

Since the circumference of a circle is $2\pi r$, the circumference of this circle is:

$$2\pi \cdot 3\sqrt{2} \quad \text{or}$$

$$6\sqrt{2}\,\pi \quad \text{(Looks like answer \textbf{E}.)}$$

Did you notice?	Once again they've given you just enough information. By telling you the hypotenuse of the triangle, you can figure out the other legs. But only if you remember that they're also the radius. And only if you notice from the diagram that ABC is a right triangle. And of course, $3\sqrt{2}\pi$ is one of the choices. Don't stop until you've reached the answer!

Long, repeating patterns are predictable.
You can make quick jumps to the end of the line.

Doughnuts are placed, one at a time, into a series of seven bags. The first doughnut goes into bag number one, the second doughnut goes into bag number two, the third into bag number three, and so on. The pattern is repeated until 86 doughnuts have been distributed. Into which bag will the 86th doughnut be placed?

(A) Bag number two
(B) Bag number three
(C) Bag number four
(D) Bag number five
(E) Bag number six

I LOVE MATH!

What do they want you to do?
They want you to draw seven bags and count out 86 doughnuts. Better still, they'd love it if you left your test and went to a doughnut shop to solve this problem. (Actually, that would work pretty well, but you'd waste an hour. You'd also spend a lot of money, because I'll tell you right now, the owner of Donald's Donuts is not about to let you borrow his merchandise. And on top of everything, you'd come back covered with raspberry jelly and powered sugar, and I hate to even think what that would do to your answer sheet.) You're not going to do that, are you? Well, what are you going to do?

Let's try a similar example. "Today is Monday. What day will it be exactly 86 days from now?"
Here's the way to handle it. There are seven days in a week. So every seventh day will also be a Monday. In other words, every multiple of seven — seven days from now, fourteen, twenty-one, and so on — will be a Monday. Now, what's the highest multiple of seven in 86? Right, 84. That means the 84th day will be a Monday. The 85th day will be a Tuesday. And I bet you've already figured out the 86th day.

Back to the original problem. Do you see that it's the same question? Except we're dealing with doughnuts instead of days. Every seventh doughnut will go into bag number seven. Which means the 84th doughnut will go into bag number seven. Doughnut 85 will be dropped into bag number one. And the 86th doughnut will be placed into bag number two — answer **A**.

If you came up with the right approach to this question, congratulations! You're really beginning to think like a master SAT-taker, and you should reward yourself! In fact, have a doughnut. Or two.

Did you notice?

You arrive at the correct answer in much less time by using this method. The secret is finding the connection between the ridiculous question and the math you already know. If you ever find yourself embarking on a long counting journey while taking the SAT, the best thing to do is stop, pull back, and figure out what the pattern is. Then look for the shortcut to the end of the pattern, and the answer.

42 When multiplying two numbers with the same base but different exponents, keep the base number and ADD the exponents.

Which of the following is equal to $3^4 \cdot 3^3$?

(A) 3^7
(B) 3^{12}
(C) 6^7
(D) 6^{12}
(E) 9^{12}

This confuses a lot of people — which is why those sneaks in SAT-land keep using it on their tests. But actually, this is one of the easiest operations in all of math. All you have to do is add two numbers (and they're such little numbers, too.)

If you're multiplying two numbers with the same base, just keep that base and add the exponents. And you're done.

So in the above example, $3^4 \cdot 3^3 = 3^7$. **(A)**

Pretty simple when you're sitting there, eating a Twinkie, with your feet up on the kitchen table, isn't it? The trick is to remember the rule. Judging by the similarity of the answers, they're guessing you won't.

Now a couple of words of caution. If the base numbers are NOT the same, you can't take this easy shortcut. When multiplying $2^3 \cdot 3^2$, you have to figure out what each value is, then multiply:

$2^3 = 2 \cdot 2 \cdot 2 = 8$, and $3^2 = 3 \cdot 3 = 9$, so $2^3 \cdot 3^2 = 8 \cdot 9 = 72$.

Another rule: If you have a number with an exponent, and that entire expression has an exponent, you keep the base and MULTIPLY the exponents. So:

$(5^3)^4 = 5^{12}$

And one more thing: get your feet off the kitchen table.

> **The Rules**
>
> $3^4 \cdot 3^3 = 3^7$
> $X^y \cdot X^z = X^{y+z}$
>
> $2^3 \cdot 3^2 = 8 \cdot 9 = 72$
> $X^y \cdot Y^z = X^y \cdot Y^z$
>
> $(5^3)^4 = 5^{12}$
> $(X^y)^z = X^{yz}$

Did you notice? The answers accommodate every mistake you could make with this question. Whether you choose to multiply the exponents, add the bases, multiply both, or add both, there's an incorrect answer just for you. To remember the rule, use $2^3 \cdot 2^2$. Expand it and you'll see that it equals $(2 \cdot 2 \cdot 2) \times (2 \cdot 2)$, or $2 \cdot 2 \cdot 2 \cdot 2 \cdot 2$, or 2^5. On the other hand $(2^3)^2 = (2 \cdot 2 \cdot 2) \times (2 \cdot 2 \cdot 2)$, or 2^6.

The length of one side of a triangle CANNOT be greater than or even equal to the sum of the lengths of the other two sides.

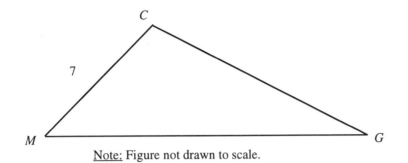

Note: Figure not drawn to scale.

In triangle *MCG* above, the length of *CG* is twice the length of *MC*. Which of the following CANNOT be the length of *MG*?

(A) 10
(B) 14
(C) 15
(D) 18
(E) 21

See how the word CANNOT is in capital letters in the question? When they do that, it means they're really serious. They don't just mean it may not, or possibly would not, or even conceivably should not. They mean it CANNOT. Never!

What this rule of triangles is saying is that, in any triangle (including the ones right below here),

$a + b$ will always be greater than c

and

$a + c$ will always be greater than b

and

$b + c$ will always be greater than a.

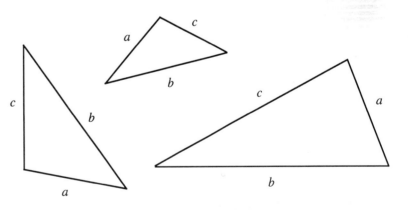

In the problem above, we're told *MC* is 7, and we're told *CG* is twice that, or 14. Therefore, according to the eternal law of geometry we have just mastered, the length of *MG* cannot — excuse me, CANNOT — be greater than or even equal to 7 + 14, or 21. The answer, then, is **E** -- the length *MG* could not be.

Did you notice?

The original triangle above is drawn in such a way that it looks as though MG must be longer than CG. Do not base your answer on that observation. The SAT is almost never interested in your perception of an illustration, but rather your understanding of a mathematical law. In this case, the word CANNOT should lead you to think about which laws deal with the lengths of sides of triangles.

44

The old 3-4-5 right triangle is a regular guest on the SAT. Learn to recognize it before you've done a lot of unnecessary work.

An ostrich runs 300 kilometers north, then turns right and runs 400 kilometers east. How many kilometers will the ostrich be from its starting point?

(A) 300
(B) 350
(C) 400
(D) 500
(E) 700

You've seen this before. It's a variation on a classic SAT question. Usually, it's a car driving north 30 miles, then turning west and going 40 miles, or a plane flying south 400 miles, then east 300 miles. Or maybe it's a person running east 3 miles, then north 4 miles. Whatever. They're all based on the same thing: the Pythagorean Theorem, a scary-sounding rule that is really very easy to learn.

The Pythagorean (Easy-to-Learn) Theorem
In any right triangle, when you take the lengths of the two legs, square them, and add those squared lengths together, the sum will be equal to the square of the length of the hypotenuse. In algebraic terms, we say $a^2 + b^2 = c^2$. (You know, it doesn't sound so easy when you say it like that. Let's look at a few examples.)

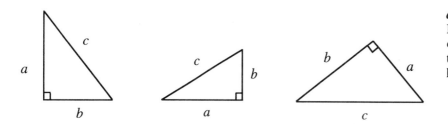

$a^2 + b^2 = c^2$
In each right triangle, the squares of the two shorter legs, added together, equal the square of the hypotenuse (the longest leg).

There are a few special right triangles that are referred to by their lengths. There's the 6-8-10 right triangle. And the 5-12-13 right triangle. But the most famous right triangle of all, the one that gets mobbed at the beach and can't go out for dinner or a movie without being hounded for autographs, is the 3-4-5 right triangle.

If you're told that the legs of a right triangle measure 3 and 4, you know instantly that the hypotenuse is 5. What's even more exciting is the fact that this formula works for the multiples of the three numbers. So there's a whole family of aunts, uncles, and cousins of the 3-4-5 right triangle. There's 6-8-10. And 21-28-35. And 30-40-50. And 300-400-500. And so on.

| **Did you notice?** | We didn't really finish the question at the top of this page. If the ostrich runs 300 kilometers north and then 400 kilometers east, we're looking at a gigantic 3-4-5 right triangle. (We know it's a right triangle because the turn from north to east is 90 degrees.) So the answer, without doing any math, is 500 kilometers (D). I hope you also noticed I used the metric system, which is preferred by most ostriches. |

45 Reduce the possibility of a mathematical error by simplifying the given information as much as possible before solving for the variable.

If $80 \cdot 10^n = 8 \cdot 10{,}000$, then $n =$

(A) 2
(B) 3
(C) 4
(D) 5
(E) 6

This problem is designed more to slow you down and make you waste time, but the chance of a careless mistake or just general confusion is clearly there also. The key to this one is to simplify before you do anything else:

$$80 \cdot 10^n = 8 \cdot 10{,}000$$

$$= 80{,}000$$

Now we see that 80 times $10^n = 80{,}000$. Therefore, 10^n must equal 1,000. And if you know anything about exponents, you know that $1{,}000 = 10 \times 10 \times 10 = 10^3$. So $n = 3$ (answer **B**).

An alternative approach is to see that, in going from the left side of the equation to the right side, the 80 has been divided by 10, giving us the 8. In order for the equation to remain balanced, the 10^n must have been <u>multiplied</u> by 10 to give us 10,000. So:

$$10 \cdot 10^n = 10{,}000$$

$$10^n \quad = 1{,}000$$

$$n \quad = 3$$

(Answer B again.)

Did you notice? The first approach is more arithmetic-based and easier to set up. The second requires some inside-the-head thinking and a little logic. The second approach also creates the possibility of dividing when you should multiply, which will produce one of the incorrect answers. (The danger is that when you come up with an answer and that answer is one of the choices, you tend to think it's right.)

The area of a parallelogram is _height_ times _base_.
Remember: the height is the line perpendicular to the base.

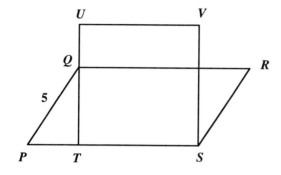

Note: Figure not drawn to scale.

What is the area of parallelogram _PQRS_ above, if _QR_ is three times the length of _PQ_ and the area of square _STUV_ is 144?

(A) 6 (B) 30 (C) 40 (D) 60 (E) 75

This is a dangerous question, because your stubborn brain wants to multiply the length of _PQ_ times the length of _QR_ and accept that as the area of the parallelogram -- which it is not. The area we're looking for can be found with the formula: A = height x base.

In this case, the base is _PS_ (or _QR_), and the height is _QT_. We can easily figure out the base, because they've told us it's three times the length of _PQ_. So the base is 15. Once we've found the height, we'll multiply it by 15 and have our answer. But the problem is, how do we find the height?

If we could find the length of _PT_, which is the base of right triangle _PQT_, we'd be able to determine the length of _QT_, because we know the hypotenuse (_PQ_) is 5. But we can't know for sure what _PT_ is -- not yet. First we have to look at that square, which isn't sitting there just to take up space.

They've given us the area of the square, which is a sneaky, SAT-way of telling us the length of the square without actually telling us. If the area is 144, the length of one side of the square, say _ST_, is 12.

Since we already know the length of _PS_ is 15, and now we know _ST_ is 12, we can be sure the length of _PT_ is 15 minus 12, or 3. Now look again at right triangle _PQT_. The hypotenuse is 5 and _PT_ is 3. Using the Pythagorean Theorem, we quickly determine that the length of _QT_ is 4. Therefore, the area of parallelogram _PQRS_ is 4 times 15, or 60 (answer **D**).

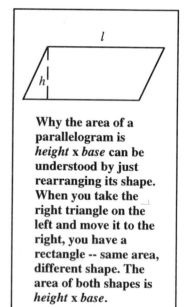

Why the area of a parallelogram is _height_ x _base_ can be understood by just rearranging its shape. When you take the right triangle on the left and move it to the right, you have a rectangle -- same area, different shape. The area of both shapes is _height_ x _base_.

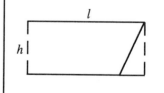

Did you notice?

If you were to multiply 5 times 15, you'd get an area of 75. Does it surprise you that 75 is one of the answer choices? It shouldn't. You can arrive at all of the incorrect answers through careless mistakes or faulty logic. The figure is not drawn to scale -- don't let it throw you. And with parallelograms, when we say base, we're talking about the length. So area = _h_ x _b_ or _h_ x _l_. Same thing.

47 If the question involves a wheel making a certain number of complete revolutions, it's usually about circumference.

A truck's right front wheel has a radius of 2 feet. How far along a straight and level road has the center of the wheel traveled when the wheel has made 5 complete turns?

(A) 4π
(B) 10π
(C) 20π
(D) 25π
(E) 35π

First thing to do? Draw a picture! It's just a circle with a radius of 2, like this:

What's the point of this question? It's that each time a circle makes one complete turn, it "travels" a distance equal to its circumference. (Imagine breaking the circle at some point and then opening it up to form a straight line. The length of that line would be the same as the circumference -- it's also the distance the center point would travel as the circle makes one complete revolution. In fact, you can think of that line as the center point's path.)

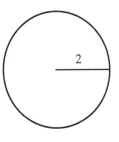

So all we need is the formula for circumference. It's:

$$C = 2\pi r$$

$$= 2\pi(2)$$

$$= 4\pi$$

Now be careful! The answer you just got, 4π, is the distance traveled by the center of the wheel after *one* complete turn. (And, of course, it's one of the answer choices.) However, the question is asking about *five* complete turns. So we need to do one more step:

$$5 \cdot 4\pi = 20\pi \quad \text{(Answer } \textbf{C}\text{)}$$

Did you notice? A couple of things. First, when they throw in a little extra information, such as the car's "right front wheel," don't get sidetracked. No matter which wheel we're dealing with, the facts (and the answer) will be the same. Also, remember that the center point moves with a turning circle, and travels the same distance (in a straight line) as any point on the circle itself.

48 A clock that runs fast is ahead of (later than) the correct time. A clock that "loses" minutes per hour will be behind (earlier than) the correct time.

A defective clock runs 10 minutes fast every hour. If the defective clock is reset at 9 a.m., what will the correct time be when the clock says 3 p.m. that same day?

(A) 1:15   (B) 1:55   (C) 2:00
(D) 2:05   (E) 4:00

A FEW MORE HOURS OF THIS AND I'LL BE ABLE TO TAKE A WHOLE DAY OFF!

This question was obviously dreamed up by someone who spends a lot of time in high schools. How would you approach this one? I would either draw a chart or a bunch of clocks. A chart would be faster, I think, and less confusing.

Let's take it one step at a time. At 9 o'clock, this defective clock is set to read the correct time. But it zips through a whole hour in just 50 minutes. That means at 9:50, it will say 10:00. After just 50 more minutes, at 10:40, the clock will read 11:00. And so on. Now we'll make that chart.

Correct Time	Defective Clock Time
9:00	9:00
9:50	10:00
10:40	11:00
11:30	12:00
12:20	1:00
1:10	2:00
2:00	3:00

As you can see, we keep adding 50 minutes to the correct time, while adding another hour onto the defective clock time. That reflects the fact that the clock runs "10 minutes fast every hour." So solving this problem isn't a matter of calculation. It's a matter of finding a pattern and then repeating it until we've reached where we want to go.

The correct answer is **C**.

Why is this question so difficult? I think it's because, once again, the SAT writers have taken advantage of another little quirk in the English language. When we say a person is running fast or running ahead, they get where they're going *earlier* than expected. But when a clock runs fast, or is running ahead, it shows a time that is *later* than expected. Just as bewildering, a person who's running behind arrives *later* than he should, but when a clock runs slow, or behind, it shows a time that is *earlier*. Now you know why I never wear a watch. Remember: you're only as confused as you think you are.

Did you notice?

We didn't calculate anything. We could have: the clock is advancing 6 hours, and because it gains 10 minutes per hour, it will gain 60 minutes in the 6 hours. But the whole thing reminded me of the fence post question, and that made me a little nervous (those answers bunched around 2:00 didn't help). So I took the safe route. Which is okay, because I didn't waste a lot of time and I got the right answer.

When translating words into an equation, just write what it says. The word *is* or *are* represents the equal sign (=).

The number of accounting majors, *A*, at a certain college is 55 more than twice the number of math majors, *M*. This information is best expressed by which of the following?

(A) $A = 2(M \times 55)$
(B) $2A = M \div 55$
(C) $M = 2A + 55$
(D) $A = 2M + 55$
(E) $2M = 55 - A$

This is an SAT favorite. It's easy, but seems to be confusing for a lot of people. I think the reason is that the answers look so similar to each other. If you stare at them for more than a couple of seconds, your brain turns into a shredded wheat kind of substance, which is wonderful with sliced peaches but doesn't handle algebra very well at all.

See what it says at the top of the page? At the risk of being repetitive, just write what it says.

The number of accounting majors, *A*, at a certain college is 55 more than twice the number of math majors, *M*.

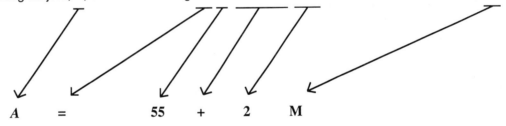

$$A \quad = \quad 55 \quad + \quad 2 \quad M$$

The confusing part of this question is probably the phrase "55 more than twice the number of." To simplify, break it down into smaller pieces. The number of math majors is *M*. Twice the number of math majors is 2*M*. And 55 more than 2*M* is (55 + 2*M*), or (2*M* + 55). So:

$$A \quad = \quad 2M + 55 \quad \text{(Answer D)}$$

 **Did you notice?** | The answers look alike, so the best approach is to ignore them at first. Work out what you think the equation should look like, then find the one that most closely matches it. If you begin with the answers, you may have to test each one, which could waste time (especially if the answer is D or E). Also remember, "55 more than" means addition, while "55 times as many" would mean multiplication.

In a triangle, the larger the angle, the larger the side opposite that angle.

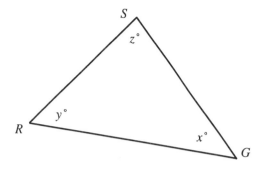

In △RSG above, $z > y > x$ and $SG = 12$. Which of the following could NOT be the length of RS?

(A) 5   (B) 6   (C) 8   (D) 10   (E) 14

What is this question about? It's about the relationship between the measures of the angles of a triangle and the lengths of the sides of that triangle. As the angle opens up (gets larger), the length of the line needed to form a triangle also gets larger. Look:

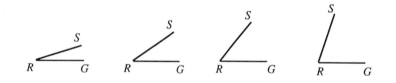

Do you see that as the angle gets bigger, the distance from *S* to *G* also gets bigger? (Of course you do: you're looking right at it.) So when comparing two angles in any triangle, the side opposite the larger angle is larger than the side opposite the smaller angle. If the angles are equal, the opposite sides are equal.

In the problem above, ∠G is the smallest angle, because *x* is less than both *y* and *z*. So *RS* has to be the shortest side (it's opposite the smallest angle). Since *SG* is 12, *RS* must be less than 12. Therefore, *RS* could not equal 14 (answer **E**).

| 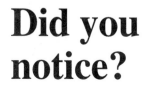 **Did you notice?** | Some of the information is given in the diagram, and some is given in the problem itself. This is typical of the SAT -- just another little, tiny, irritating way to confuse you. Also, be extra alert on any question that asks you to find the answer that could NOT be correct. Your brain is not used to thinking that way, and tends to head for answers that COULD be correct. (Don't you hate being a grownup?) |

51

When dealing with percents, *of* means
multiplication. And *x*% of *y* equals *y*% of *x*.

For
instructions on
questions 51-70,
see the last page
of the
Introduction.

Column A	Column B
976% of 12	12% of 976

It should be clear by now what they would love for you to do with this question. It should be equally clear that you shouldn't do it.

They want you to do the arithmetic. Even with a calculator, this is going to take more of your time than it deserves. Some of you will realize this immediately, and try to guess which one is bigger. You will employ some fairly sophisticated logic. Many will conclude that A is bigger. An equal number will be sure B is bigger. All will be wrong. Look carefully at the values in the two boxes:

976% of 12	**12% of 976**
=	=
976 x 1/100 x 12	**12 x 1/100 x 976**

Do you now see that the two values are exactly the same? By doing it this way, there's no need to calculate what the two values actually are. All you want to know is, which one is bigger, or are they equal? As long as the numbers (and the operations) involved are the same, the answer will be **C**.

There are a couple of variations on this question to watch out for. Here's one:

97% of 400	40% of 970
97 x 1/100 x (40 x 10)	**40 x 1/100 x (97 x 10)**

Equal again. How about this one?

90% of 42	40% of 92

Do you see how this one is different? In this case, A is larger. Be careful!

Did you notice?	There are no variables (*a*'s, *q*'s, or *x*'s) in this question. All of the values are expressed in normal, everyday numbers. They are clear and definite. That means the answer cannot be D. If you're only dealing with numbers, then it must be possible to compare them. Either they're equal, or one is larger than the other. (For a different twist on this idea, see question 52.)

52

If at least two different answers are possible, the answer must be D.

<table>
<tr><th>Column A</th><th>Column B</th></tr>
<tr><td>10% of one of the following: {90, 100, 200, 250}</td><td>20% of one of the following: {90, 100, 200, 250}</td></tr>
</table>

This is a completely different question from number 51. True, there are only numbers involved. But there are several possible values on each side. To be precise, there are four possible values for A, and four for B. Here's what they look like.

10% of 90 = **9**	20% of 90 = **18**
10% of 100 = **10**	20% of 100 = **20**
10% of 200 = **20**	20% of 200 = **40**
10% of 250 = **25**	20% of 250 = **50**

Depending on which values you select, A could be larger than B (for example, 25 is larger than 18), B could be larger than A (50 is larger than 10), or they could be equal (20). Once you've determined that there are at least two possible answers to a quantitative comparison question, the answer must be **D**.

There are variations on this question to watch out for. Which answer would you select for this one?

<table>
<tr><th>Column A</th><th>Column B</th></tr>
<tr><td>10% of one of the following: {50, 60, 70, 80}</td><td>30% of one of the following: {30, 40, 50, 60}</td></tr>
</table>

Again, there are four possible values for A, and four possible values for B. However, in this case, ALL of the values for B are larger than ALL of the values for A. So no matter what, B must be the correct answer.

Did you notice?

In many of these questions, the answer could be A, B, or C. But your goal is not to find an answer it could be. You're looking for the answer it has to be. If more than one answer is possible, you must choose D. In the question near the bottom of the page, the higher numbers on the left are there to throw you off. The fact is, the largest A could be is 8, while the smallest B could be is 9.

Familiar information can be presented in a way that automatically draws you toward the wrong answer.

Column A	**Column B**
Circumference of a circle with diameter x.	Circumference of a circle with radius $2x$.

Do you think the answer to this question is C? If so, you have fallen victim to a typical SAT trick.

If the question had asked you to compare a circle with diameter $2x$ and a circle with radius x, the answer would be C, because the circles would be equal. But read the question again. If it helps, draw a quick picture.

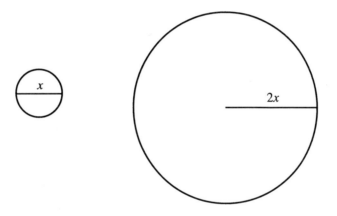

See? A circle with radius $2x$ is much bigger than a circle with diameter x. So the answer is **B**.

If it helps, plug in numbers. Let x equal 3. Then the circle under column A would have a diameter of 3, and the circumference would be 3π. (Circumference = πd, or $2\pi r$.) The circle under column B would have a radius of 6, and the circumference would be 12π. The answer is still B.

Did you notice?	The confusion caused by this question has to do with the fact that the diameter of a circle equals two times the radius. So when you see diameter x under column A and radius $2x$ under column B, your well-trained mind automatically jumps to the careless conclusion that the circles are equal. It's sneaky, it's nasty, it's unpleasant. Welcome to SAT-land.

54

When looking for a pair of numbers to meet two different requirements, approximate first, then adjust to get the exact numbers. Or use algebra.

<u>Column A</u>   <u>Column B</u>

The sum of two positive integers is 30.
The difference of these integers is 12.

| The smaller of the two integers. | 8 |

The two sentences above are crucial to solving this problem, but they must be taken together. Either piece of information without the other is useless. Let's take a look at them.

The sum of two positive integers is 30. That means the integers could be 1 and 29, 2 and 28, 3 and 27, and so on. But now we add a new requirement: *The difference of these integers is 12.* That means that when we subtract one of the integers from the other, we get 12. There's only one pair of numbers that fits both restrictions: 9 and 21. They add up to 30 and their difference is 12.

How did we find the two integers? Trial and error. Educated guessing. Start somewhere that makes sense: 10 and 20, for example. They add up to 30 and their difference is 10. Then adjust each slightly.

Or we can use algebra. If the smaller number is x, the larger is $x + 12$. The sum of the two is 30. So:

$$x + (x + 12) = 30$$

$$2x + 12 = 30$$

$$2x = 18$$

$$x = 9 \quad \text{If the smaller number is 9, the larger number is } (9 + 12), \text{ or 21.}$$

Back to the question. Which is larger, the smaller of the two integers or 8? Since the smaller of the two integers is 9, and 9 is larger than 8, the answer is A.

Here's an alternate approach that works just as well. Let's assume the smaller of the two integers *is* 8. Because the difference of the two numbers is 12, the other integer must be 20. But the sum of 8 and 20 is only 28. The first requirement said the sum must be 30, so the numbers must be larger than 8 and 20. In other words, the smaller number must be larger than 8. No matter what that number is, then, A is the correct answer.

| **Did you notice?** | The larger of the two answers (A) has the word *smaller* in it. Is that meant to confuse you? Probably. Make sure you know the definitions of words like *sum* and *difference*. Not knowing them would get in the way of finding the correct answer to many SAT questions. Finally, when looking for relatively small numbers, such as two whose sum is 30, the trial-and-error method is often faster than algebra. |

Read the question quickly, then analyze the chart or graph so that you understand what it's saying, then read the question again. (Part II)

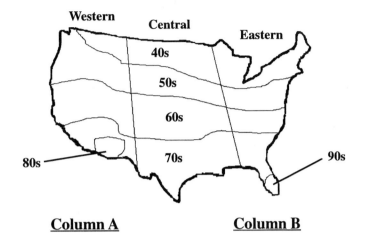

Column A   **Column B**

The above map shows temperature bands across three regions of the United States.

The highest temperature in the Western region.	92°

I know you won't have trouble with this. I just liked the map.

    The key to this question -- and all SAT questions involving maps, graphs, charts, or tables -- is to understand the information presented. Most often, it's easy stuff. They're just testing to see whether you can translate data from one medium to another (for example, map to words).

    Don't be intimidated by what may at first appear to be some complex meteorological map. It's just showing the temperatures in different parts of the country (nothing you haven't seen on the six o'clock news).

You're being asked to compare the highest temperature in the Western region (A) with 92 degrees (B).

    If you understand the map at all, you know that the warmest area in the Western region is in the 80s. It's that little round area in the Southwest. Obviously, if the highest temperature in the Western region is in the 80s, then 92 must be higher. So the answer is **B**.

Did you notice?	The highest temperature on the entire map is in the 90s. Yet the specific number, 92, is presented as the value for B. This is done to confuse you into thinking the answer might be D, since you can't know if a temperature "in the 90s" is higher or lower than 92. But be careful: it's in the 90s in southern Florida, and the question is addressing only the Western region.

56 Often you can use the information given to determine what something does *not* equal, even though there's no way to determine what it *does* equal.

<u>Column A</u>   <u>Column B</u>

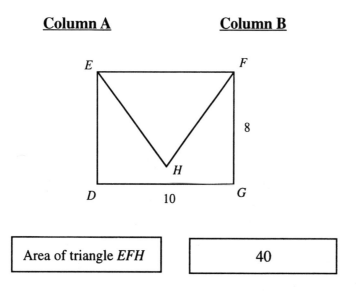

| Area of triangle *EFH* | 40 |

What's the area of a triangle? They give you the formula! It's one-half the base times the height, or 1/2(*b* x *h*). But wait, you protest, they didn't tell us the dimensions of the triangle! True. However, what they have told us is quite enough.

Look first at the answers. Where did that 40 come from? The area of the rectangle is 8 x 10, or 80. The length of the rectangle, 10, is also the base of the triangle. So the *b* in the formula for area of a triangle is 10. Now if (IF) point *H* were sitting on line *DG*, the height of the triangle would be the same as the width of the rectangle, or 8. See that? IF. In that case, what would the area of the triangle be?

1/2 (10 x 8), or 40.

So that's where the 40 came from. The area of the triangle *would be* 40 IF the height were 8. But the height is less than 8, because the triangle is not as tall as the rectangle. So the area of triangle *EFH* must be less than 40, and the answer must be **B**.

Did you notice? We never did figure out what the area of that triangle is. And the truth is, we never will, because there's just too much to do as it is and we don't have time for hopeless cases. They didn't give us enough information to determine the area of that triangle. But they gave us just enough to be sure it had to be less than 40. And in a quantitative comparison question, that's all we need to know.

57 Combine what you know about the angles of a triangle and supplementary angles (in each case, they add up to 180). And be sure to select the answer you've found!

<u>Column A</u>   <u>Column B</u>

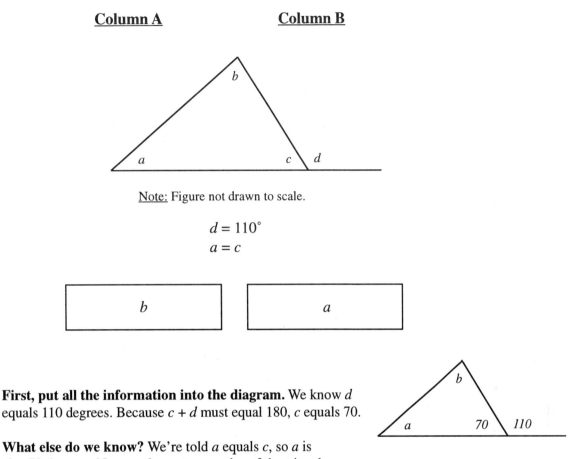

Note: Figure not drawn to scale.

$$d = 110°$$
$$a = c$$

b	a

First, put all the information into the diagram. We know d equals 110 degrees. Because $c + d$ must equal 180, c equals 70.

What else do we know? We're told a equals c, so a is also 70 degrees. Now we know two angles of the triangle. Remember that the three angles of the triangle add up to 180 degrees. Because $(a + c)$ equals $70 + 70$, or 140, b must equal $180 - 140$, or 40.

If $a = 70$ and $b = 40$, then a is bigger.

But watch out! Notice they've put a under column B and b under column A. The answer is not A, it's a -- which is under column **B**. (Close call there.) If it helps, put your finger right onto the correct answer and say to yourself, almost out loud, "The answer is B."

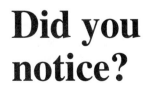

Did you notice?

There's an exterior angle in the diagram. Most people don't remember this rule, but the exterior angle (in this case d) equals the sum of the two opposite angles (in this case a and b). It doesn't save you any time in this problem, because you can't find a until you find c. But make sure you know the rule. It will come in handy. Also, b looks bigger than a, but notice the figure is not drawn to scale.

58

When doing a probability question, make sure you're working with the correct numbers.

PICK A DIGIT, ANY DIGIT

<u>Column A</u>	<u>Column B</u>

A set of twenty cards is numbered from 1 to 20. There is exactly one number on each card and no two cards have the same number. A card is drawn at random from the deck.

The probability that the card will have a 2-digit number.	The probability that the card will have a 1-digit number.

It should be absolutely obvious to you which answer those people over at SAT headquarters are praying you'll choose for this question. They have set you up like a crooked street con, just salivating at the thought that you'll take their bait.

The automatic response, the one we all instinctively give, (and the one that will cost us points from our score), is C. Aren't they equal? Aren't there ten 1-digit numbers and ten 2-digit numbers? No. There are nine 1-digit numbers and eleven 2-digit numbers. The numbers 1 through 9 are the 1-digit numbers. The 2-digit numbers consist of those from 10 to 20. Go ahead, count them up. There are eleven.

The answer, then, must be A, because there are eleven 2-digit numbers, and only nine 1-digit numbers. The probability of selecting a 2-digit numbered card is 11/20, while the probability of drawing a 1-digit numbered card is 9/20. But why does this happen? Why do we want to say they're equal? And how do the SAT sneaks know we want to say they're equal?

It happens because we read the question and, in our minds, change it. We change it to read,

The probability that the card will have a number from 1 to 10.	The probability that the card will have a number from 11 to 20.

This, of course, is a completely different question, and we can all see that now, can't we? (The answer to this new version would be C.) Remember, SAT questions and answers have been well-tested before you ever see them. The testmakers know which traps work best.

Watch out for this question, or some variation of it, on your own SAT.

Did you notice?

In the answers, they put the probability for a 2-digit number on the left (column A) and the probability for a 1-digit number on the right (column B). A minor point, perhaps, but just another little thing that can throw you off. For some reason, we expect the 1 to be on the left, and the 2 to be on the right. Which is why they often don't do it that way.

59 Unless defined otherwise, a variable can be positive or negative, fraction or whole number.

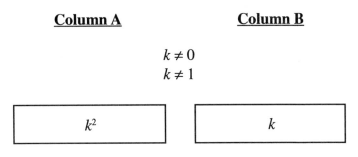

Column A   **Column B**

$$k \neq 0$$
$$k \neq 1$$

k^2	k

Once again, it's easy to see what they're hoping you'll put for this one. They want you to go through this line of thinking:

If k doesn't equal zero, then it must be either a positive or negative number. And because squaring a number makes it bigger, and because the square of a negative number is a positive number, k^2 has to be bigger than k. Of course the number 1 is an exception. If you square 1 you still have 1, and they would be equal. But they've told us k doesn't equal 1 either. So the answer must be A.

Not so fast. It's true that for any positive whole number k greater than 1, the square of k is greater than k. And it's true that the square of any negative number is a positive number, and a positive number is always bigger than a negative number.

But, what if k is a fraction less than 1? What would happen if k equaled 1/2, for example? When you square 1/2 you get 1/4, which is smaller. In fact, when you square any fraction less than 1, you end up with an even smaller fraction. And since the information provided doesn't exclude fractions less than 1, there are many values for k that will give us an answer of B. If A is possible and B is possible, the answer must be **D**.

Remember, when presented with a question that asks you to compare the values of defined variables, go through all the possibilities:

What happens if the variable equals 0?
What happens if it's negative?
What happens if it equals 1?
What happens if it equals 2?
What happens if it's a fraction?

Did you notice?

This question presents a minimum of information, making it appear to be a simple problem. In two short statements, they eliminate the possibility that k could be negative, 0, or 1. And by saying it doesn't equal either of those whole numbers, they get you thinking in terms of only whole numbers. What they're counting on is that you won't ask the question, "What if k is a fraction?" But, of course, you will.

60 A larger perimeter doesn't necessarily mean a larger area.

<u>Column A</u>	<u>Column B</u>
The area of a rectangle with perimeter 24	The area of a rectangle with perimeter 26

Once more, the choice of the uninitiated, uninformed, and unprepared will be answer B. It just seems that the rectangle with the larger perimeter will have the larger area. The key is that it could -- but it doesn't have to. Let's take a look.

The secret, if you haven't seen this question before, is to draw a few pictures. (Not one, not twelve: *a few*.) Could the answer be B? Sure:

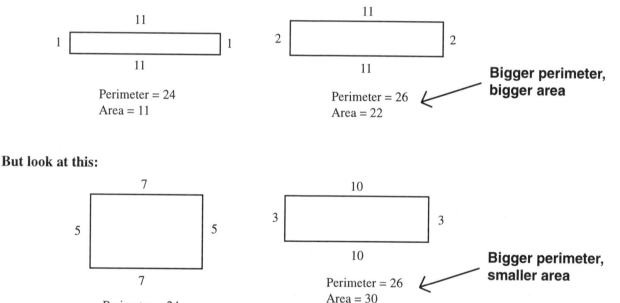

So what valuable lesson have we learned here? That when comparing two rectangles, the one with the larger perimeter doesn't necessarily have the larger area. Either one could be larger, or they could be equal. So the answer is **D**.

# Did you notice?	The area of a rectangle depends on the shape. If the rectangle is long and narrow, it has a small area. As the rectangle approaches the shape of a square, the area grows to its largest size. For example, in the rectangle with perimeter 24, a 6 x 6 square would have the largest area, 36. Important: If you're looking for the largest *possible* area of two rectangles, it will always be the one with the larger perimeter.

61 Think about what they tell you -- and what they don't tell you.
There is a tendency to sometimes assume the wrong thing.

Two line segments, *AB* and *CD*, are parallel and have
exactly the same length. *X* is the midpoint of *AB*.

<u>**Column A**</u>   <u>**Column B**</u>

Length of *XC*	Length of *XD*

This is one of my all-time favorite SAT questions, reprinted here with only the slightest of modifications (with any luck, this book will never be popular enough to attract the attention of the copyright lawyers).

It is a perfect trap -- seemingly harmless, innocent, and so easy! Show this question to a hundred people, and very nearly a hundred will answer it the same way. Let me give you a small hint on how to approach it: I found it as number 24 out of 35 questions. What does that tell you? It should tell you that it's going to require a little work, or at least a little thinking. So what are you going to do? Go ahead, draw a picture and come up with an answer.

Is this what you drew?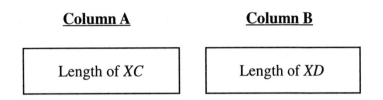

Maybe you drew vertical lines, but I'd be willing to bet my last number-2 pencil that your interpretation of this question looks something like this. Guess what. So did mine! The first time I answered this question and checked the answer, I was convinced they had made a mistake in the book. I had fallen into the trap of relying on appearances.

When you draw the lines this way, *XC* and *XD* are the same length. So one possible answer is C. But it isn't the *only* answer, because it isn't the only way to draw the lines. The fact is, *XC* could be longer, or *XD* could be longer. So the answer has to be **D**. Here, look for yourself:

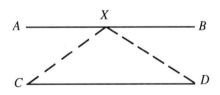

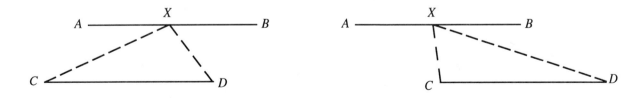

Did you notice? When we hear that two lines are "parallel and have the same length," we automatically draw them one right above the other, or one right beside the other. It doesn't occur to us that one may be slid all the way to the left or right. Why do we do this? Maybe because we crave order and long for a perfect world. That doesn't make us bad people, but it does create a slight disadvantage on the SAT. Careful!

62 Use small numbers whenever possible. Make a chart.
Be sure you're answering the question!

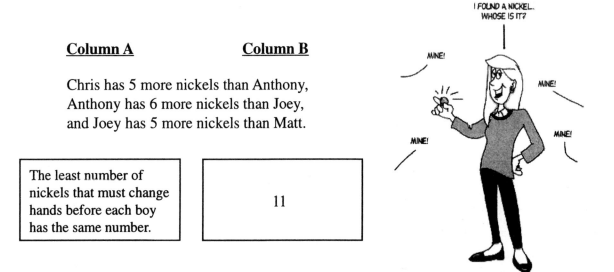

<u>Column A</u>	<u>Column B</u>

Chris has 5 more nickels than Anthony,
Anthony has 6 more nickels than Joey,
and Joey has 5 more nickels than Matt.

The least number of nickels that must change hands before each boy has the same number.	11

As with most SAT questions, this one looks a lot scarier than it is.

Let's begin by deciding that, because the question is asking for the least number, we should use the smallest numbers possible. The least number of nickels someone can have is 0, and Matt has the least number, so let's give him 0 nickels. From here we can make a simple chart:

M	0
J	5
A	11
C	16

(No need to write out the complete names -- that wastes time!)

Now we know how many nickels each boy starts with. But how many should each end up with? Adding the nickels, we find there are a total of 32. These are to be divided evenly among the four boys. So each will end up with 8 nickels. From here it's just a matter of distributing the coins in the most efficient way. Let's do this:

M	0		M	0 + 8		M	8
J	5	$\longrightarrow$	J	5 + 3	$\longrightarrow$	J	8
A	8 + 3		A	8		A	8
C	8 + 8		C	8		C	8

With Chris giving 8 of his nickels to Matt, and Anthony giving 3 of his nickels to Joey, the boys all end up with the same number. A total of 11 nickels changed hands, and the answer is **C**. And everybody's happy. Except Chris and Anthony.

Did you notice?	We could have used higher numbers -- such as 10, 15, 21, and 26. With 72 nickels to be divided among four boys, each would have ended up with 18. Again, Chris would give Matt 8 and Anthony would give Joey 3. The answer will always be 11. But by using the smallest numbers, we reduced the risk of making a mistake. (Another danger: if you thought the answer was 8, you would have put B.)

Sometimes the ability to visualize changes to a diagram will get you to the answer quickly, and with less confusion.

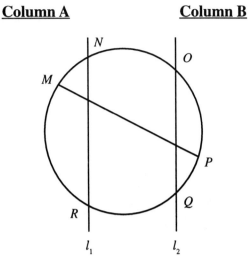

Column A	Column B

Note: Figure not drawn to scale.

$$l_1 \parallel l_2$$

MP is a chord of the circle.

Length of arc *MNO*	Length of arc *PQR*

Your brain looks at this problem and thinks the following:

"Those two lines are parallel. That other line, *MP*, is zagging across the two parallel lines, which means those two angles are equal. And so the arcs opposite those angles must be equal."

That's not a far-fetched assumption, but it's wrong.

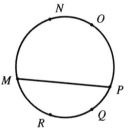

It says *MP is a chord of the circle*, and it also says *Figure not drawn to scale*. That means they've drawn it the way they want you to see it (the arcs look equal), but it can be drawn other ways. The fact is, *MP* can be any chord on the circle. Look at the two diagrams to the right. (The parallel lines have been removed to make things clearer.) Now imagine holding point *P* down with your finger and sliding point *M* down the circle. See how arc *MNO* gets bigger, while arc *PQR* stays the same? The arcs could be equal, but they don't have to be. The answer is **D**.

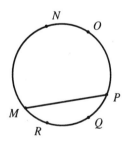

Did you notice?

The old "Figure not drawn to scale" plays a key role. It means things can move, and if things can move, other things may get bigger or smaller. The placement of the parallel lines is an effective diversion. We're conditioned to respond to them, especially when they're intersected by another line. The corresponding angles that result are equal, but as we've seen, that doesn't mean the arcs must be equal.

64 Don't keep re-reading a confusing word problem.
Translate it into clear and simple equations.

<u>Column A</u>   <u>Column B</u>

If Allison had 12 more bottles of hairspray, she
would have 5 times as many as she actually has.
If Meaghan had 2 fewer bottles of hairspray,
she would have half as many as she actually has.

The number of bottles of hairspray Allison actually has.	The number of bottles of hairspray Meaghan actually has.

Once again, a simple problem disguised in confusing language. Let's translate each sentence into an algebraic expression, and you'll see how simple this hairspray situation really is.

Let's let *A* equal the number of bottles Allison actually has. Then *M* should represent Meaghan's hairspray collection. Here's how it all translates:

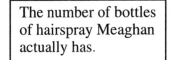

If Allison had 12 more bottles of hairspray, she would have 5 times as many as she actually has.

$$A + 12 = 5A$$

If Meaghan had 2 fewer bottles of hairspray, she would have half as many as she actually has.

$$M - 2 = 1/2M$$

Now let's just solve the two equations:

$$M - 2 = 1/2M$$

$A + 12 = 5A$	$2M - 4 = M$
$12 = 4A$	$2M = M + 4$
$3 = A$	$M = 4$

So Meaghan has 4 bottles of
hairspray, Allison 3. The correct
answer is **B**.

Did you notice? | One of the equations involves addition and multiplication. The other involves subtraction and division (or multiplication by a fraction). This is no accident, but just another way the SAT will try to drive your mind into a state of mild hysteria. Don't let them succeed. Translate each sentence into a simple equation. Then solve each equation separately.

65 They don't think you know what words like *remainder* mean. Make sure you do.

Column A	Column B
The remainder when 47,950 is divided by 5.	The remainder when 20,242 is divided by 7.

This is just the kind of SAT question that causes many people to go home after the test with an inflated sense of confidence. Then the scores arrive and the balloon bursts.

If you answered A for this question, you fell into the trap.

What does your mind want to do immediately? Right, divide the numbers. Especially since you have that incredibly powerful calculator starving for its next meal. Well, put a muzzle on that thing. This is not a division problem! They're not asking you what the result is when you divide those numbers. If they were, you could just look at them and choose A. You'd be dividing a bigger number (47,950 is a lot bigger than 20,242) by a smaller number (5 is smaller than 7). So of course the result would be a bigger answer for the one on the left. But let's read the question again. They want the remainder. Important math vocabulary word. The remainder is not the quotient. When you divide 2 into 9, you get 4 with 1 left over. That 1 is the remainder!

Back to the question. Under column A, can you see that 5 will divide beautifully into 47,950? Any number that ends in a 0 is divisible by 10, and any number divisible by 10 is also divisible by 5. In other words, the remainder for column A is 0.

Under column B, when you divide 7 into 20,242, it doesn't go in evenly. You end up with a remainder of 5. So even though the *quotient* under column A is larger, the *remainder* is smaller. And the answer is **B**.

Did you notice? Not only can you get the wrong answer here, but you can waste some time, as well. Your calculator will give you a decimal answer, when you're looking for a whole number remainder. So you'll probably end up dividing column B again in long division (remember that?) just to make sure. Or, you'll fall for the primary trap, see that the quotient for column A is larger, and choose that. (I'm praying you won't.)

Short questions with several confusing parts should be taken one piece at a time.

I CAN'T TELL YOU HOW MUCH I HATE PERCENTS.

<u>Column A</u>	<u>Column B</u>

w percent of 20 is less than 8.

$2w$	70

A letter, a percent, a less than, a number out of nowhere, two times the letter. The whole picture is overwhelming. Your eyes move from one part of the question to the other in a clockwise motion. You're getting very sleepy...

WAKE UP! This is the SAT. You don't have time to waste, and you certainly don't have time to sleep. Let's push this question around a little, then answer it and move on. One piece at a time. First, we'll translate the words into a mathematical statement:

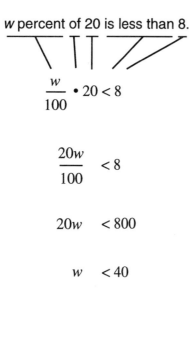

w percent of 20 is less than 8.

$$\frac{w}{100} \cdot 20 < 8$$

$$\frac{20w}{100} < 8$$

$$20w < 800$$

$$w < 40$$

So what have we learned from this little ordeal? Only that *w* is less than 40. Could we have done it more quickly? Sure. If you know that 8 is 40 percent of 20, you would know immediately that *w* must be less than 40 percent. In either case, if *w* is less tha 40, then $2w$ must be less than 80.

So under column A we have some number that's less than 80 and under column B we have 70. Could A be bigger? Yes, it could be any number greater than 70 and less than 80. For example, if *w* equals 39, then $2w$ would equal 78. Could B be bigger? Yes, if *w* equaled 10, then $2w$ would be 20. In that case, A would be less than 70. Could they be equal? Absolutely, because *w* could equal 35, and $2w$ would be 70.

The answer, then, is **D**.

# Did you notice?	That *less than* sign never changed. That's because all we did was multiply and divide with positive numbers. We treated it just like an *equal* sign and it treated us just like a friend. The key here is activity: Don't get hypnotized by the question, staring and re-reading until you're having an out-of-body-experience. Get in there and do something.

Don't just blindly pick the winner. Wisely select the correct answer.

<u>Column A</u>	<u>Column B</u>
The time it takes a car to travel 65 miles at an average speed of 60 miles per hour	The time it takes a car to travel 60 miles at an average speed of 65 miles per hour

If you had to answer this question without giving it any thought, what would you put? Most people say B. Why? Because the mind is drawn toward the faster car. If it were a race, we would naturally focus on the winner, and car B is traveling a shorter distance at a faster speed. So it will get there first.

Unfortunately, B is not the correct answer. The question is asking for the time it would take to travel the given distance at the given speed. The slower car requires *more time*, therefore that value is *greater*.

To put it into concrete numbers, let's use the formula for rate, time, and distance:

Rate x Time = Distance

For car A, then,

$60 \times$ **Time** $= 65$

Time $= 65/60$ **hours**

For car B,

$65 \times$ **Time** $= 60$

Time $= 60/65$ **hours**

Car A's time is 65 minutes.

Car B's time is about 55 minutes.

Because car A has the "larger" time, the correct answer is **A**.

Did you notice? They use the same number in both parts of the answer -- they just reverse their order. Did this confuse you? Don't let it. Just stick with your logic. One car is traveling slower and over a greater distance. The other car is traveling faster and for a shorter distance. The slower car will take *more time* than the faster car. If you just think it through, you will steer clear of the traps.

68 If you see a number raised to some gigantic exponent (anything bigger than 5), don't do the calculation. There's a faster way.

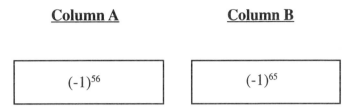

Column A	Column B
$(-1)^{56}$	$(-1)^{65}$

Just a reminder before we begin: the answer to this question cannot be D. It involves only numbers and, clearly, the value of each column can be determined. Okay, now that we have that straightened out, let's get started.

Actually, there isn't much to do with this one. Both sides have a negative 1 being raised to a certain power. So the value for each column will be either positive 1 or negative 1, depending on the exponent.

Confused? Here, look:

$(-1)^1 = -1$
$(-1)^2 = 1$
$(-1)^3 = -1$ (Each time you raise the exponent, you're multiplying
$(-1)^4 = 1$ the *previous* value by another negative 1. So the sign
$(-1)^5 = -1$ just keeps switching back and forth.)
$(-1)^6 = 1$

And on and on, forever. Negative 1 raised to an *even* exponent gives you a positive 1. Negative 1 raised to an *odd* exponent brings you back to negative 1. So:

$(-1)^{56} = 1$
$(-1)^{65} = -1$

Positive 1 is greater than -1, therefore, $(-1)^{56}$ is greater than $(-1)^{65}$, and the answer is **A**.

Did you notice?	You could have spent a couple of days raising negative 1 to the 56th and 65th powers, but that would probably not be a good use of your time. It's much better to just know that no matter what the exponent, if the base is 1 or -1, the result will be either 1 or -1. The tendency, if you don't know that, is to choose the answer with the higher exponent. (Good logic, maybe, but bad math.)

69 When you begin with an average and then introduce
a higher number into the mix, the average increases.

The average (arithmetic mean) of x and y is 12.

<u>**Column A**</u>   <u>**Column B**</u>

| The average of x and y | The average of x, y, and 15 |

Don't do anything yet. Just think. If your average in Social Studies is 82 and you get a 90 on the next test, what happens to your average? It goes up. You know this instinctively. That's exactly what's happening here. You've taken two tests and your average is 12. On the third test, you get a 15. The result is a higher average (not quite enough to make the honor roll, but a definite increase).

Can't get comfortable with this logical approach? Need proof? Okay.

If the average of x and y is 12, that means:

To find the average of x, y, and 15:

$$\frac{x + y}{2} = 12$$

$$\frac{x + y + 15}{3} = A$$

and

$$\frac{24 + 15}{3} = A$$

$$x + y = 24$$

$$\frac{39}{3} = A$$

Now go over here, because there's more room.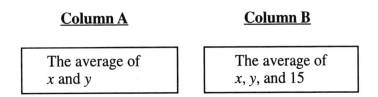

$$13 = A$$

So the average of x and y is 12, and the average of x, y, and 15 is 13. The answer is **B**.

Did you notice? This question could take you about five seconds to answer, if you trust your instinctive understanding of mathematics. If you need to work it out, it should still take you less than a minute. Be sure that when you want to find the average of three numbers, you add them and divide by 3. And as always, don't be confused by the phrase *arithmetic mean*. It just means *average*.

When faced with variables, percents, and other inconveniences, break the problem down into manageable parts.

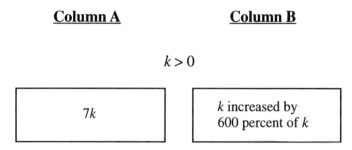

Column A	Column B
$k > 0$	
$7k$	k increased by 600 percent of k

Before you race out of the room, terrorized by the very sight of this question, consider the fact that the College Board (the people behind your favorite standardized test) consider this a 5 on a 1-5 scale of difficulty. In other words, this is as hard as they say it gets. What does that mean? It means if you can master this question, you are capable of mastering any question in this book -- and any question you're likely to encounter on the SAT.*

But it has letters, and *greater than* signs, and percents! So what? All you have to do is understand what they're really saying in that box under column B. Right? Isn't that the problem? Okay, let's do it.

k increased by 600 percent of k

Let's begin with 600 percent of something. No, wait, let's begin with 100 percent. What would 100 percent of k be? Right: k. What would 200 percent of k be? $2k$. Starting to see a pattern? What would 600 percent of k equal? Correct again: $6k$.

Now the tricky part. What do they mean by k increased by 600 percent of k? We just determined that 600 percent of k is $6k$, remember that? Okay, so the real problem, the root of all evil here, is the simple phrase "increased by." Do we really know what that means? We need to, because we're trying to determine k increased by $6k$. Let's take a brief side-step.

What would "k increased by 3" equal? Right, $k + 3$. What would k increased by $10m$ equal? Right again: $k + 10m$.

If we just follow this through, k increased by $6k$ must equal $k + 6k$, or $7k$. Which means that k increased by 600 percent of k equals $7k$, and the correct answer is **C**.

*Results based on average 29 years' prep time. Your scores may differ. See store for details. Void where prohibited by law. Some settling of contents may occur during shipping. Tax, title, and other fees not included. Offer subject to change without notice.

 **Did you notice?** They tell you that k is greater than 0. Is that a necessary piece of information, or something thrown in just to slow you down? If k were equal to 0, the answer would be C. And if k were, say, -5, the answer would be C. Assuming this were the last question in the section, and you didn't have a whole lot of time left, that little $k > 0$ could cause you to doubt yourself and leave the answer blank.

71

The word MUST means it's true in all cases.
The word COULD means it's true in at least one case.

If x is an integer greater than zero, which of the following MUST be an integer?

I. x^2

II. $\dfrac{2x+1}{3}$

III. $\dfrac{20}{4x}$

(A) I only
(B) II only
(C) III only
(D) II and III only
(E) I, II, and III

Remember, an integer is just a whole number, like 1 or 23 or 854. Clearly, if x is an integer, then x^2 must also be an integer. There's no way to square a whole number and turn it into something other than another whole number. So the answer must be either A or E. Do you see why? (If statement I is true, then the answer must contain statement I. Only A and E do.)

Does II have to be true? No, just plug in 2 for x and you get 5/3, which is never going to make it as an integer. How about III? If we plug in 2 for x, we get 20/8, which is also not an integer. And since MUST means always, II and III are not true because in each case we found at least one value for x that does not produce an integer.

Only I MUST be true, so the answer MUST be A.

What if the question had said, which of the following COULD be an integer? Then we'd only have to find one example that worked for each to be considered true. We already know statement I could be an integer (in fact, it MUST be an integer). Is there a value of x that would make statement II an integer? How about 4? That would give us 3, which is an integer, last time anybody checked. And for statement III, if x equals 1 or 5, the result is an integer. So the answer to this alternate question — which of the following COULD be an integer? — would be E.

Did you notice?	We are conditioned to solve for the variable every time we see one. So when confronted with the three statements above, we immediately want to go to work finding values for x that will make the statements true. And as we've seen, those values for x exist. But that's not the point. They want to know which statements are true no matter what x equals. Only statement I meets that requirement.

To calculate how many different combinations are possible, first determine the number of possibilities for each part, then multiply those numbers together.

How many different 3-digit numbers are possible
if the 1st and 3rd digits cannot be 1 or 0?

(A) 26
(B) 300
(C) 640
(D) 810
(E) 1,000

If you even had a thought that you might just list all the different combinations, forget it. You'll miss your next birthday, Thanksgiving, and the first three nights of Hanukkah.
　　Here's the only way to solve this problem.

First, understand the information given. They want 3-digit numbers. The 3-digit numbers cannot be repeated (so you can't count 287 twice). The first digit cannot be 1 or 0. The 3rd digit cannot be 1 or 0.

Second, understand the question they're asking. They want to know how many different 3-digit numbers are possible, according to the rules we just repeated.

Third, solve the problem. Begin by drawing three dashes to represent the 3-digit numbers:

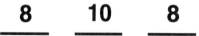

On each of those three dashes, write the number of different possible digits that can go in that place. Remember, there are ten digits, including 0. The first place cannot contain a 1 or 0, so it has eight possible digits. The second digit can be any of the ten, so write a 10 there. The third digit cannot be 1 or 0, so write an 8 on the third dash. You should have this:

8　　10　　8

Now just multiply the three numbers: 8 x 10 x 8 = 640, and we're finished.
The answer is **C**.

| **Did you notice?** | If you were to make the mistake of adding the numbers (8 + 10 + 8), you would get 26, which is one of the answer choices. If you put a 10 on each dash and multiplied them, you would get 1,000, which is also one of the answers. And if you were to use 9 • 10 • 9, you would get answer D. There are many ways to make a mistake, but there's certainly no reason to. This is an easy question. |

73

Cancel out everything you can before multiplying fractions.

$$\frac{3}{5} \cdot \frac{5}{7} \cdot \frac{7}{9} \cdot \frac{9}{11} \cdot \frac{11}{13} =$$

(A) $\frac{3}{13}$ (B) $\frac{7}{13}$ (C) $\frac{7}{9}$ (D) $\frac{11}{13}$ (E) $\frac{34}{45}$

This is a frequently-used question on the SAT. They will most likely change the numbers, but the technique you need to solve it remains the same.

First, be sure you understand that those dots between the fractions represent multiplication. (I know you knew that, but there's no harm in making sure, is there?)

Second, recognize what they're hoping you'll do with this question. They would love for you to actually multiply 3 times 5 times 7 times 9 times 11 for the numerator, then multiply 5 times 7 times 9 times 11 times 13 for the denominator. Even if you're using a calculator, you'll probably go through it twice to check your answer. And when you end up with 10,395 over 45,045, and see that there's no such answer choice, you'll have to reduce the fraction to its lowest terms. All this will take way too much time. Even if you get the right answer, the SAT will have won the battle, because you will have spent more than a minute on a question that deserves about ten seconds of your life.

Here's what to do. Wherever you see a number in one of the numerators that matches a number in one of the denominators, cross them both out. (They equal 1, which is just taking up space.) This is what you should end up with:

$$\frac{3}{\cancel{5}} \cdot \frac{\cancel{5}}{\cancel{7}} \cdot \frac{\cancel{7}}{\cancel{9}} \cdot \frac{\cancel{9}}{\cancel{11}} \cdot \frac{\cancel{11}}{13} =$$

Whatever isn't crossed off is your answer. And can you believe it? It matches one of the choices! (Answer A) But if this method makes you nervous, or if you suspect it's just a cheap trick and not really math, then look at it this way:

$$\frac{3}{5} \cdot \frac{5}{7} \cdot \frac{7}{9} \cdot \frac{9}{11} \cdot \frac{11}{13} = \frac{3 \times 5 \times 7 \times 9 \times 11}{5 \times 7 \times 9 \times 11 \times 13} = \frac{5 \times 7 \times 9 \times 11 \times 3}{5 \times 7 \times 9 \times 11 \times 13} = \frac{5}{5} \cdot \frac{7}{7} \cdot \frac{9}{9} \cdot \frac{11}{11} \cdot \frac{3}{13}$$

$$= \frac{3}{13}$$

Did you notice?	The matching numbers are placed right next to each other, so they're easy to recognize. Obviously, the question has been designed so that you can solve it in a number of ways. But the best way is equally obvious: by canceling out matching numerator and denominator values, you save time, do less work, and greatly reduce the risk of making an error.

74 The volume of a rectangular solid, including a cube, equals *length* times *width* times *height*.

A room in the shape of a rectangular solid has 84 square feet of floor space. If the volume of the room is 756 cubic feet, what is the height of the room, in feet?

(A) 7   (B) 8   (C) 9   (D) 12   (E) 16

This is one of the easiest SAT questions! All you have to remember is that the volume of a rectangular solid is length x width x height. Actually, you don't even have to remember it -- they give you the formula on the first page of each math section. Just remember to use it.

With most rectangular solid problems, it helps to draw a quick diagram. It isn't really necessary in this case, though, once you realize that the area of the floor (or ceiling) divided into the volume will give you the height. In fact, on the SAT, all you have to do is plug what they give you into the formula to find what they want.

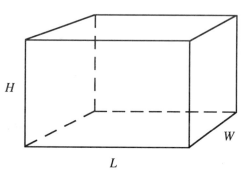

$$L \times W \times H = \text{Volume}$$

$$L \times W = \frac{\text{Volume}}{H}$$

$$H = \frac{\text{Volume}}{L \times W}$$

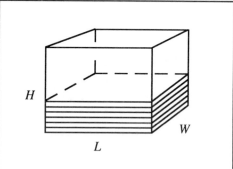

They've given us the volume of the room, 756. And because the area of the floor ($L \times W$) is 84, when we divide 756 by 84, we get the height of the room:

$$H = \frac{756}{84}$$

$$= 9 \quad \text{(Answer C)}$$

Think of a rectangular solid as having a carpeted floor and being completely filled with a stack of carpets, all the size of the floor. The volume is just the area of the bottom carpet ($L \times W$) times the number of carpets in the stack (H).

Did you notice?   It doesn't matter which sides you label as the length, width, and height of the rectangular solid. The point is, you multiply all three dimensions to get the volume. If you divide the volume by any one of the three dimensions, you will get the other two. And if you divide the volume by any two of the dimensions, you will get the third dimension. (Don't spend a lot of time drawing the diagram!)

75 Confusing word problems can usually be turned into simple algebraic equations -- which can then be solved and checked.

The local dealership had a two-day sale on all cars in stock. On the first day, they sold 1/3 of their cars. After 4 vehicles were sold on the second day, 1/2 of the original number of cars were still on the lot. How many cars were in stock before the sale began?

(A) 8   (B) 12   (C) 24   (D) 36   (E) 48

Headache hazard! Just because a question is written as a word problem doesn't mean your brain can deal with it in that form. The fact is, you could reread the above problem many times and never see it more clearly. (You could probably walk to the dealership and count the cars faster than you could make sense of this question.) Let's simplify.

Algebra to the rescue! Believe it or not, algebra can be your friend. You can use it to translate confusing words into confusing symbols. So what's the point? The point is, you can keep changing the words in the original problem and it doesn't get you anywhere. But you can simplify an algebraic equation until you've found what you're looking for.

Let's begin by assigning a variable to represent the number we're trying to find.

x = **the number of cars originally in stock.** There are two ways to go from here.

1/3 of x cars were sold on the first day, or **$x/3$**.

On the second day, 4 more cars were sold. Now, $x/3 + 4$ cars have been sold. The problem also tells us that at that point, 1/2 of the original number of cars had been sold. Therefore:

$$x/3 + 4 = x/2$$

$$2x + 24 = 3x$$

$$24 = x$$

2/3 of x cars were left after the first day, or **$2x/3$**.

On the second day, 4 more cars were sold. Now, $(2x/3) - 4$ cars were left. The problem also tells us that at that point, 1/2 of the original cars had been sold, so 1/2 are left. Therefore:

$$(2x/3) - 4 = x/2$$

$$4x - 24 = 3x$$

$$x = 24$$

 **Did you notice?**

One approach addresses the number of cars *sold*. The other talks about the number of cars *left*. Because 1/2 of the cars were eventually sold, $x/2$ doesn't change -- half sold means half left! However, depending on which approach you choose, you must decide whether to add or subtract the 4. Either way, once you've solved for the variable, plug it back into the equation to make sure it works.

76

Fence post questions are about events, not intervals between the events. And they don't have to contain fence posts.

John is on a special diet and must drink an 8-ounce glass of water every 20 minutes, beginning at 8 a.m. If he drinks his last glass at 4 p.m., how many glasses of water will he consume that day?

(A) 22   (B) 23   (C) 24   (D) 25   (E) 26

Remember the question about the fence we saw on page 34? What we learned is that our brains want to count the intervals -- the spaces between the fence posts -- rather than the fence posts themselves. What we also learned is that this tendency will leave us holding the wrong answer.

Well, this is basically the same question. We quickly determine that John will be drinking water 3 times an hour for 8 hours. So we assume the answer is 24 and move right along to the next question, never suspecting that we have just fallen victim to another SAT act of terrorism.

As with the fence post question, the key here is that we're not counting the blocks of time between glasses of water. We're counting the glasses of water -- and there's one at the very beginning (8 a.m.) and at the very end (4 p.m.) In other words, one more than we think there should be. Here's an illustration. If you can count, you can answer this question.

8:00   8:20   8:40   9:00   9:20   9:40   10:00 10:20 10:40   11:00 11:20 11:40

12:00 12:20 12:40   1:00   1:20   1:40   2:00   2:20   2:40   3:00   3:20   3:40

That's 24, and one more at 4:00 makes 25 glasses. (Answer **D**)

4:00

Did you notice?

This question involves a very convenient interval of 20 minutes. That allows you to divide quickly into 60 minutes and get 3 glasses an hour. But what if it had been one glass every 40 minutes? Just multiply the total number of hours, 8, times 60 minutes per hour. So the total time frame consists of 480 minutes. Because 40 goes into 480 twelve times, the correct answer to that question would be... 13.

Long calculations are almost never necessary on the SAT. First, simplify wherever you can.

If $v = 12$, then $\dfrac{v^2 + 3v}{v} =$

(A)   15
(B)   17
(C)   60
(D) 147
(E) 180

Are you multiplying, adding and dividing to solve this problem? It sounds like you're working much too hard! Which is exactly what those SAT people want you to do. Because if you make one little mistake, the answer you come up with won't be there, and you'll have to start all over. (Wouldn't you just hate that?)

Here's how to simplify the situation. Before you plug in the 12, simplify the expression. A lot of those v's don't need to be there. In fact, when you factor out a v, this is what you have left:

$$\frac{v\,(v + 3)}{v}$$

Or, better yet, just:

$v + 3$

And since $v = 12$,

$v + 3 = \mathbf{15}$. Answer **A**. Done! On to the next challenge!

|  **Did you notice?** | Factoring out the v doesn't just save you time and work. It also reduces the possibility of a mistake. Instead of having to square 12, multiply 3 times 12, add those two results, and divide by 12 -- all you do is factor out the v, then add 12 plus 3. Think about it: if you did make a mistake with all that calculating, the answer you got wouldn't be one of the choices. Then what? |

Vertical angles are equal. Corresponding angles are equal.
An exterior angle equals the sum of the two opposite angles.

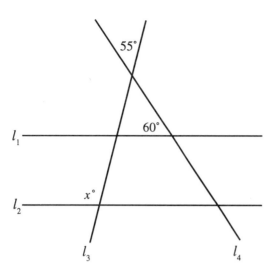

In the figure above, lines l_1 and l_2 are parallel, and are intersected by lines l_3 and l_4. What is the measure of x?

(A) 55   (B) 60   (C) 115   (D) 125   (E) 130

Without even reading the question, just looking at the diagram, you should immediately realize this has to do with parallel lines, vertical angles, corresponding angles, and the angles of a triangle. So call that stuff out of your mental toolbox and let's go to work.

     This is another of those questions that you can solve about nine different ways. But which way is the fastest, and the cleanest? We'll work backwards, beginning where we want to end up: we're looking for x. Which angle, if we knew its measure, would tell us the measure of x?

How about this one?

     Do you see that this angle is equal to x? (They are corresponding angles.) It also happens to be an exterior angle of that small triangle. The exterior angle equals the sum of the two opposite angles. One of those opposite angles is 60. The other, the top angle of that small triangle, is equal to the 55 degree angle (they are vertical angles). So the sum of the two opposite angles is 60 + 55, or 115.

     The exterior angle equals 115°, and so does x. (Answer C)

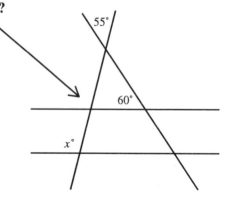

Did you notice? There's at least one more good way to solve this one. The top angle of the small triangle is 55. So the two known angles of that small triangle add up to 115. That means the third angle of that triangle is 65 (they total 180). Because l_1 and l_2 are parallel, the angle in the lower left corner of the large triangle also equals 65. That angle, plus x, must equal 180. So x equals 115. Did you follow that?

79

When looking for rate, time, or distance, always begin with the formula (R x $T = D$) and rearrange the variables to solve for the one you need.

Jackie and Jim are driving at a constant speed of z miles per hour. How many hours will it take them to travel 800 miles?

(A) $800 - z$

(B) $800z$

(C) $800 + z$

(D) $\dfrac{z}{800}$

(E) $\dfrac{800}{z}$

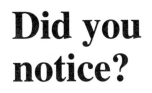

I WAS NEVER VERY GOOD AT ALGEBRA, BUT AREN'T YOU DRIVING A LITTLE FAST?

I want you to see how they give you everything you need. They will most likely give you the formula, but even if they don't, you already know it. I'll prove it. If you were traveling at 50 miles per hour for 2 hours, how far would you have traveled? Right, 100 miles. How did you do that? You multiplied the rate, 50, times the time, 2, to determine the distance, 100. That's exactly the formula you need to solve this problem.

Rate x Time = Distance   (R x $T = D$)

In this case, they've given you the speed. They call it z. And they've given you the distance: 800 miles. Let's plug those two things into the formula:

(z) x $T = 800$

Now we're just one step away from the answer. Because we're looking for T, we need to move everything else to the other side of the equal sign. To do that, we'll divide both sides by z.

$$T = \dfrac{800}{z}$$

The correct answer is **E**.

| **Did you notice?** | You're not solving for z. We never do find out what z equals (although the way Jim drives, I bet it's pretty fast). But they're not asking us to determine what z equals, or what T equals. What they want is the time, T, expressed in terms of the rate, z, and the distance, 800. All of the answers contain both values, which is why you would have a hard time without that formula. Make sure you know it! |

80

The "mixture" problems can be solved algebraically, or with plain old logic.

A mixture of candy is made by blending chocolate-covered raisins, at $4 a pound, with chocolate-covered almonds, at $9 a pound. If the resulting mixture is worth $7 a pound, how many pounds of the almond candy are needed to make 10 pounds of the mixture?

(A) 2   (B) 3.5   (C) 4   (D) 5   (E) 6

There are two approaches: algebra and logic. Take your pick.

Algebra Approach

Let x equal the number of pounds of the chocolate-covered almonds needed for the mixture. Then $10 - x$ will equal the number of pounds of the chocolate-covered raisins needed. Since the cost of the almond candy is $9 a pound and we need x pounds, the total cost of the almond portion of the mixture will be $9x$ dollars. Meanwhile, the raisin candy costs $4 dollars a pound and we need $(10 - x)$ pounds, so the total cost of the raisin portion will be $4(10 - x)$ dollars. The final mixture costs $7 a pound, or 70 dollars for 10 pounds. Therefore:

$9x + 4(10 - x) = 70$

$9x + 40 - 4x = 70$

$5x + 40 = 70$

$5x = 30$

$x = 6$ (Answer E)

Logic Approach

As you add more of the cheaper candy to the mixture, the price per pound drops. As you add more of the higher-priced candy, the price per pound increases. If you were to make a mixture with an equal amount of each candy, that mixture would cost $6.50 a pound. Why? Because $6.50 is exactly halfway between $4 and $9.

Since the mixture in the problem is $7 a pound -- which is more than $6.50 -- then more than half the mixture must consist of the almond candy. The total mixture is 10 pounds. If the almond candy represents more than half the mixture, there must be more than 5 pounds of it.

The only answer choice greater than 5 is (**E**).

Did you notice?

One approach isn't better than the other. With the algebra approach, you can plug 6 back into the original equation, figure out that there will be 4 pounds of the cheaper candy, and make sure everything works. However, there is always the risk of making a mistake with the arithmetic and coming up with an answer that isn't there. With the logic approach, you end up with the only possible answer.

Figure not drawn to scale? Re-draw it, and label it properly!

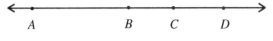

Note: Figure not drawn to scale.

If *C* is the midpoint of segment *AD*, and if *AB* = 2
and *CD* = 8, what is the length of *BD*?

(A) 16   (B) 14   (C) 10   (D) 8   (E) 2

Another one of those irritating SAT questions that you try to keep reading and re-reading, because you think if you stare at it long enough, the answer will come to you. It won't. You have to re-draw the picture and plug in the information. Why make your brain work harder than it has to? They tell you *C* is the mid-point. Why didn't they draw it that way? I don't know. But you'd better do it. And if *AB* is 2 and *CD* is 8, shouldn't *CD* look longer than *AB*? Try this:

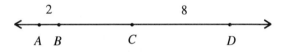

There, isn't that refreshing? And we know more stuff, too. We know that *C* is the midpoint of *AD*. Doesn't that mean *AC* equals *CD*? Of course it does. So if *CD* is 8, then *AC* is also 8. And, if *AC* is 8 and the *AB* part of it is 2, then the *BC* part must be 6. And when we put that juicy bit of information into the diagram and re-check what they're asking, we're left with a fairly simple addition problem:

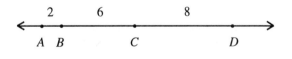

BD = BC + CD

BD = 6 + 8 = 14 (Answer **B**)

Did you notice? | They tell you *C* is the midpoint of *AD* -- but they draw it to look as though *B* is the midpoint. Then all the information about the other lengths either goes into your head wrong, or it just seems confusing. Instead of working out the problem, your mind is wasting time trying to reconcile what it sees with what it's told. Take a few seconds to give yourself a diagram that makes sense.

When counting different possibilities, make a quick chart. It might take a little longer, but you'll end up with the right answer!

$$3, \quad 2, \quad 0, \quad 6, \quad 0, \quad 8$$

If a and b represent any two unequal numbers from the list above, how many <u>different</u> values are possible for ab?

(A) Six   (B) Seven   (C) Eight   (D) Nine   (E) Ten

If you're trying to do this in your head, forget it. There are too many places where you could get confused, and the answers don't leave any room for guesswork or error. What we need is a chart.

But before we make a chart, let's make sure we understand the question. Here's how I interpret this one:

When you multiply two different numbers from the list, you get a certain product. How many different products are possible? Is that what you thought they were asking? Good, let's make the chart.

a	3	3	3	3	2	2	2	2	0	6	6	6	8
b	2	0	6	8	3	0	6	8	2	2	3	8	3
ab	6	0	18	24	6	0	12	16	0	12	18	48	24

I count seven different products: 6, 0, 18, 24, 12, 16, and 48. Make sure you adhere to the requirements specified in the question. The two numbers, a and b, must be <u>unequal</u>. That means 3 times 3, 2 times 2, 6 times 6, and 8 times 8 are not allowed.

Very important: ab represents the product of a and b. It does *not* represent a 2-digit number consisting of a in the ten's place and b in the one's place. So results such as 23, 30, 62, and 83 are not valid. If that were the correct interpretation of the question, the answer would be sixteen, which is not even there.

The correct answer is **B**.

Did you notice?

I didn't bother to list every possible combination in the chart. Once I realized I already had three products of 0, and needed only one, I stopped using 0. And when I saw that all the products with 8 and another number were already in the chart, I stopped using 8. Also, notice the question uses numerals, while the answers have the numbers spelled out as words. Very yicky.

83 The principles involved in determining average and percent don't change when you combine them into one question.

On day 1, Kati and three online friends each post an average of 5 messages. How many total messages would the group have to send the next day in order for the day 2 average to show a 60% increase over the day 1 average?

(A) 8   (B) 12   (C) 20   (D) 30   (E) 32

This question has several monsters, and they're not even hiding. The first is the number 3, written out, because there are actually 4 people involved. Next, it's about computers and that "online" stuff, which intimidates some people. And it combines average and percent, two somewhat scary topics on their own. So where do we start? We've already started. We've determined that we're dealing with 4 people. That will be critical when we have to calculate the average. Which we'll do right now. Five messages a day, on average, posted by each of four people. So how many messages, total, on day 1?

$$A = \frac{\text{Total messages}}{\text{Number of people}}$$

$$5 = \frac{\text{Total messages}}{4}$$

$$20 = \text{Total messages}$$

So on the first day, the group sent a total of 20 messages. The question is, how many will they have to send the next day so that the day 2 average will be 60% higher than the day 1 average? Well, 60% of 5 is 3. If they raise the daily average by 3, their average for day 2 will be 8. (This is the key to the whole question, so if you didn't follow that, go back and read it again.)

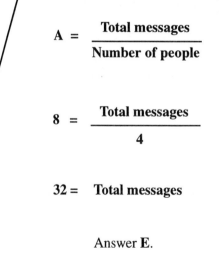

We have the new average, 8. And we still have 4 people. All we have to do now is plug this information into the formula, and we'll get the message total for day 2.

$$A = \frac{\text{Total messages}}{\text{Number of people}}$$

$$8 = \frac{\text{Total messages}}{4}$$

$$32 = \text{Total messages}$$

Answer **E**.

Did you notice?	You know how to do average and you know how to do percent. The key to this question, and almost all SAT questions, is the way it's worded. Specifically, what is being increased by 60%, and what does that mean? To show a 60% increase over a number is to determine what 60% of the number is, then add that to the original number. So 60% of 100 is 60. But a 60% increase over 100 is 160.

84

In questions involving statements that *must* be true, you can often eliminate more than one answer at a time.

There are 16 married couples taking a cruise together. For this particular group of travelers, which of the following statements must be true?

I.   At least two of these couples got married in the same month of the year.
II. At least two of these couples got married on the same day of the week.
III. At least two of these couples got married on a Saturday.

(A) I only   (B) III only   (C) I and II only   (D) II and III only   (E) I, II, and III

So there are these sixteen fun-loving couples on this big boat. Is it true that at least two of these couples must have gotten married on a Saturday? No, of course not. It's quite possible that *none* of them got married on a Saturday. In fact, maybe they all got married on a Tuesday. It could be a club of some kind.

What have we accomplished so far? More than you think. We've determined that statement III is not acceptable. Which means we can eliminate any answer choice that contains statement III: specifically, answers B, D, and E. If this were an actual test, you would be wise to cross off those three answers in your question booklet. (Think of it: you're down to two answers and all you've done was figure out that these people *could* have all gotten married on a Tuesday.)

Look at the two remaining answers. Do you notice anything? They both contain statement I. That means statement I must be true! Don't even bother to check it! Okay, check it if you want to. See? Sixteen couples, twelve months -- there has to be some duplication.

What we really need to find out is whether or not statement II must be true. And since we can very quickly recall that there are only seven days in the week, and there are still the same sixteen couples, at least two have to match. (Think about it this way: after the first seven couples tell us the day of the week on which they got married, the eighth couple has to correspond to one of the first seven.)

So statement I must be true and statement II must be true. The answer is **C**.

Did you notice?	If this question is confusing, it's because your mind is thinking in terms of probability. This is not a probability question. We're not picking two couples at random and saying *they* had to get married in the same month. We're saying at least two out of the sixteen had to. If the first twelve couples all had gotten married in different months, the remaining four would have to match somebody.

Need the area of a circle? Find the radius.

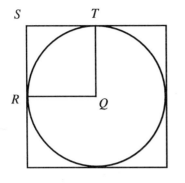

If the area of square *QRST* is 10, what is
the area of the circle with center *Q*?

(A) $\dfrac{\pi}{10}$ (B) $\dfrac{\pi}{100}$ (C) $\dfrac{5\pi}{2}$ (D) $2\sqrt{10}\,\pi$ (E) 10π

Why are they telling you the area of the little square? Because embedded in that information is
what you need to know to find the area of the circle. In other words, the radius is hidden somewhere in
this picture. If you can find it, you will be one step away from solving the problem.

Do you see it? The side of square *QRST* is also the radius of the circle. How do we know? Because
the question tells us that *Q* is the center of the circle. And the diagram shows us that points *R* and *T*,
corners of the square, are also points on the circle.
     So if we can determine the length of square *QRST*, we will know the radius of the circle. Then
we can plug the radius into the formula for area of a circle, and we'll have our answer.

The area of square *QRST* is 10. The length of a square is the square root of its area. So the length of
square *QRST* -- say *QR* -- equals the square root of 10. Which is also the radius of our circle.

Area of a circle = πr^2

$$= \pi(\sqrt{10})^2$$

$$= 10\pi \text{ (Answer E)}$$

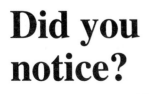

If the question says "40% less than," that's the same as "60% of."
Use the numbers 10 and 100 whenever you can in a percent problem.

If the length of square *S* is 40% less than the length of square *T*, then the area of square *S* is what percent of the area of square *T*?

(A) 16   (B) 36   (C) 40   (D) 60   (E) 84

What cute little traps have they planted into this question? Well, there are at least two visible to the naked eye. The first is the most obvious: they want you to think that if the length of a square is 40% less, then the area will be 40% less. (Not true, because you have to square the length to find the area. The squares of numbers do not have the same relationship as the original numbers. For example, 3 is half of 6, but 3^2 is one-fourth of 6^2.)

The second trap lies hiding in the words. The length of square *S* is "40% less than" the length of square *T*. That is not the same as "40% of." If square *T* has a length of 10, then "40% less than" 10 would be 6, while "40% of" 10 would be 4. See that?

Let's draw a picture to clear it up.

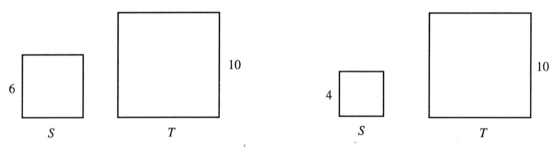

The length of *S* is 40% less than the length of *T*.   The length of *S* is 40% of the length of *T*.

Look at the pair of squares on the left. What would their areas be?

The area of square *S* is 6^2, or 36. The area of square *T* is 10^2, or 100. Since 36 is 36 percent of 100, the area of square *S* is 36 percent of the area of Square *T*. (Answer **B**)

Did you notice?

We used a length of 10 for square *T* so that the area would be 100. Then, whatever the area of square *S* turned out to be, it would be the same percent of the area of square *T*. In a percent problem, try to use 100 whenever you have the option. This question is worded to confuse. Just draw a picture and break it down. By the way, do you see why the other answer choices were used?

87 Sometimes what appears to be a confusing question involves only simple addition, subtraction, multiplication, and division.

Maria's Taxi Service charges $2.00 for the first half-mile driven, $0.75 for each of the next three miles, and $0.50 for each additional mile or partial mile after that. If the charge for a certain trip was $9.75, which of the following could have been the actual mileage driven?

(A) 8
(B) 9 1/2
(C) 11 3/4
(D) 13 3/4
(E) 19

This question makes you glad you take the bus, doesn't it? But that won't help you get the answer. As with most SAT questions, if we just break it down into simple steps, we can get to the solution with very little trouble.

We know the person paid $9.75. The first half-mile was $2.00, so let's subtract that from the total.

9.75
-2.00 (*1/2 mile*)
_____

That leaves **7.75** to account for. Each of the next three miles cost $0.75. That's $2.25. Let's subtract that.

-2.25 (*3 miles*)
_____

There's **5.50** left. How many additional miles could have been driven for that amount of money?

At $0.50 per mile, the taxi could go another 11 miles. But that would bring us to 14 1/2 miles, which isn't one of the choices. Let's try *10 miles* instead. That would account for $5.00. So far, then, the taxi has traveled 13 1/2 miles, for a charge of $9.25. An additional *quarter-mile* would be considered a partial mile, and would be charged $0.50. That would bring the total fare to $9.75 and the total mileage to 13 3/4, as in answer **D**.

Did you notice?	The use of the words "could have been" causes a certain amount of discomfort. It suggests a degree of fuzziness we don't expect in a math question. And there's a reason for that fuzziness. The fact is, the taxi could go as far as 14 1/2 miles and the charge would still be $9.75. Why? Because the last $0.50 is for anything up to the last complete mile. So a quarter-mile is charged the same as an extra mile.

Don't just solve for the variable -- answer the question. (Part III)

In Michael and MaryAnn's store, hardcover books cost $12 each and softcover books cost $8 each. If a customer paid a total of $180 and bought three times as many softcover books as hardcover, what was the <u>total</u> number of books purchased?

(A)   5
(B)   15
(C)   20
(D)   24
(E)   36

Given enough time, you could eventually figure this one out just by trying different numbers. But on the SAT, the name of the game is efficiency. So let's do it the right way. Do you remember the question about the chocolate-covered almonds? This is similar, except the books are plain. Let's look at two slightly different approaches. One involves more thinking at the end, the other at the beginning.

Approach 1

In this solution, we let x equal the number of $12 books and $3x$ equal the number of $8 books. Then $12(x)$ is the number of dollars spent on the $12 books, and $8(3x)$ is the number spent on the $8 books. The total of the purchase is $180, so:

$$12x + 8(3x) = 180$$

$$12x + 24x = 180$$

$$36x = 180$$

$$x = 5$$

Remember, x is only the number of $12 books purchased. We now have to figure out that three times as many $8 books were purchased, or 15. The total, then, is 20 (Answer C).

Approach 2

In this approach, we let x equal the total number of books purchased. So when we solve for x, we'll have our answer. Here's where the extra thinking comes in. If three times as many of the $8 books were bought, then three-fourths of the books cost $8 and one-fourth cost $12. See that? Therefore:

$$(8)\frac{3}{4}x + (12)\frac{1}{4}x = 180$$

$$\frac{24}{4}x + \frac{12}{4}x = 180$$

$$6x + 3x = 180$$

$$x = 20$$

Did you notice?

When we let x equal the number of $12 books, we got $x = 5$. That's not the answer, but of course 5 is one of the choices. Next, 15 of the $8 books were bought, and 15 is one of the choices. We have to add those two numbers to answer the question. That's what I meant by thinking at the end. In Approach 2, we had to think at the beginning, deciding that the $8 books were three-fourths of the total, and so on.

Unless otherwise stated, the square root of a number can be either positive or negative. They won't mention that, especially when it's important.

If $x^2 = 64$ and $y^2 = 49$, then the difference between the greatest possible value of $(x - y)$ and the least possible value of $(x + y)$ is

(A) 0
(B) 14
(C) 15
(D) 16
(E) 30

When it comes to the SAT, the wording's the thing. Translated, this question really says this: If you take the largest possible value of $(x - y)$ and the smallest possible value of $(x + y)$, how far apart are they?

The first thing to be aware of: x can equal 8 and -8, and y can equal 7 and -7. The SAT people are counting on your forgetting that. Or they're hoping you just won't understand the question.

As with so many SAT problems, the real key is to break it down into manageable pieces. Let's approach this question by looking at its individual parts, as restated in the paragraph above.

The largest possible value of $(x - y)$. If x equals 8 and y equals -7, subtracting them will give us 8 - (-7), or 15. That's as big as $(x - y)$ can be.

The smallest possible value of $(x + y)$. If x equals -8 and y equals -7, adding them will give us $(-8) + (-7)$, or -15, which is the least $(x + y)$ can be.

So the difference between the two values (15 and -15) is 30. (Answer **E**)

Just in case you need to see all the possibilities, here they are:

$(x - y)$	$(x + y)$
8 - 7 = 1	8 + 7 = 15
8 - (-7) = 15	8 + (-7) = 1
-8 - 7 = -15	-8 + 7 = -1
-8 - (-7) = -1	-8 + (-7) = -15

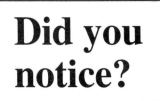

Did you notice? There's a tendency to take 15 and -15 and add them. And, of course, the result is one of the answer choices. When determining the *difference* between a positive number and a negative number, think about temperature. If the temperature goes from 15 degrees to minus 15 degrees, it drops 30 degrees. But with just a little carelessness, any of those wrong answers could look pretty good.

Total cost is price per unit times the number of units.
To convert cents to dollars, divide total cost by 100.

A grocery store sells apples for 73 cents a pound. How much, in *dollars*, would p pounds of apples cost?

(A) $73p$ (B) $\dfrac{73p}{100}$ (C) $\dfrac{100p}{73}$

(D) $173p$ (E) $730p$

There is no formula to memorize here. Just use common sense. You've bought apples before, or at least seen someone do it on television. It's not that mysterious. They charge you by the pound. If the apples cost 73 cents per pound, and you buy 1 pound, you owe somebody 1 x 73 cents, or $0.73. If you buy 8 pounds, you owe 8 x 73 cents, or $5.84.

When you buy p pounds of apples, you owe $73p$ cents. Remember, the pattern doesn't change when you switch over to letters. That's the beauty of it.

Okay, what's left? The most confusing part: changing cents to dollars! When there are letters involved!

Once again, let's figure out how to do this with numbers first. Then we'll apply the pattern to letters and before anybody knows what happened, we'll have our answer.

If we had 658 pennies, how many dollars is that? You know this: $6.58. How did we get that? We divided 658 by 100, because there are 100 pennies in each dollar. Another example. If we had 7,341 cents, how many dollars? Divide by 100: $73.41.

Here's the big one. If we had $73p$ cents, how many dollars would we have? Correct.

We'd have $\dfrac{73p}{100}$ dollars, and the answer would be **B**. Which it is.

As you have no doubt noticed by now, the answers look terrifyingly similar to each other. This will always be true in this type of question. They will take the elements of the correct answer and rearrange them into every possible combination. As you look over these various possibilities, you may experience feelings of self-doubt and confusion. This is normal. And it's what they want you to feel. Just stick to what you know. And don't let anything steer you away from what you believe to be the right answer. If you divided by 100, it's because that's what you were supposed to do. Just because one of their answers says *multiply*, that doesn't have anything to do with the math problem you just solved. Be tough.

Did you notice?	This question asked us to convert cents to dollars, and we divided by 100. Sometimes they ask you to go the other way and convert dollars to cents. In that case, *multiply* your total by 100. Again, you already know this. If you have $7.00, that's 700 cents. If you have $62x$ dollars, that's $6,200x$ cents. And so on. Once you understand what you're doing, there's no reason to be confused by letters.

Find the connections between what is known and what is being asked.

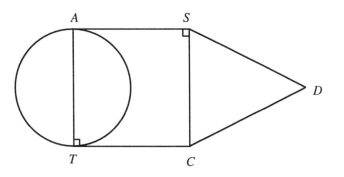

If the circumference of the circle above is 8π,
what is the perimeter of equilateral $\triangle CSD$?

What is the connection between what they've told us and what they're asking?
1. They've told us is the circumference of the circle.
2. The circumference will give us the diameter.
3. The diameter is also the side of square *TASC*.
4. The side of square *TASC* is also the side of $\triangle CSD$.
5. Triangle *CSD* is equilateral, so the perimeter is 3 times the side.

Wait just a minute, you might be saying. How do we know *TASC* is a square? See those two little boxes in the opposite corners of *TASC*? That says it has two opposite right angles, so it must at least be a rectangle. Which means *TA* = *SC*. That's all we need. (Just for the record, I know it's a square because I drew it that way.)

1. The circumference of any circle equals the diameter times π.
2. In this case the circumference is 8π, so the diameter is 8.
3. The height of square *TASC* is 8.
4. The side of $\triangle CSD$ is also 8.
5. $\triangle CSD$ is equilateral, so the perimeter is 8 + 8 + 8, or **24**.

Did you notice? There were no traps in this question. That will usually be true of these Student-Produced Response, or grid-in, questions. Most are straightforward. The question doesn't try to steer you toward a wrong answer, because no answers are provided. And while we're on that subject, remember: you are not penalized for a wrong answer on this type of question, so it doesn't make sense to leave it blank.

For any point (x,y) on a circle with center at the origin, $x^2 + y^2 = $ radius^2

A circle with center at the origin passes through the point (-5,0).
If another point on the circle is $(3,y)$, what does y equal?

First thing I'm going to do is draw that circle.
    I guess it would look something like this one:

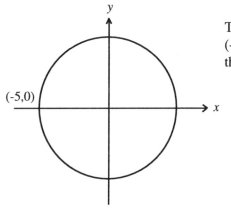

The next thing I would think about is that the point (-5,0) is on the circle. That means the distance from the origin (0,0) and that point is exactly 5. Which means the radius of the circle is 5. Next, I'd draw another radius from the origin out to about where that point $(3,y)$ ought to be.

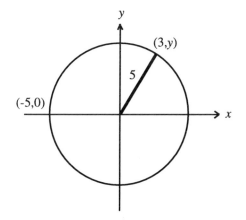

Then I'd drop a line from the mystery point right down to the x-axis. And what we'd have is a right triangle with a base of 3, a height of y, and a hypotenuse of 5. See that?

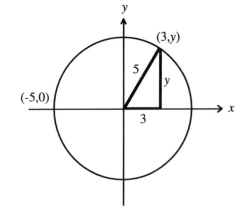

We have a right triangle, so:

$$3^2 + y^2 = 5^2 = 25$$

For any point on this circle, then, the x-value squared plus the y-value squared equals 25.

Since $3^2 = 9$, y^2 must equal 16. Then y equals 4. (It's a 3-4-5 right triangle. You saw that didn't you?) The answer is **4**.

Did you notice?   If you square the coordinates of any point on the circle, then add those squares together, that sum equals the square of the radius of the circle. Usually on the SAT, this question appears as a multiple-choice. The answer is the pair of coordinates whose squares add up to the square of the radius. To find a point that lies *outside* the circle, that sum would have to *exceed* the square of the radius.

93 Seemingly complex word problems usually translate into simple equations. The trick is to translate accurately.

CAN YOU HELP ME WITH DIFFERENTIATION OF LOGARITHMIC FUNCTIONS?

SURE, BUT FIRST I HAVE TO MAKE A WEE-WEE

In a certain high-pressure kindergarten, 1/4 of the students are enrolled in calculus, and of these, 1/3 also take advanced chemistry. If at least 3 students take both calculus and advanced chemistry, what is the LEAST number of students in the kindergarten class?

I don't know about you, but I find this question frightening. (Then again, the thought of advanced chemistry has *always* scared me.) What should we notice? Two things, I think. First, the phrase *of these*. We're told that 1/4 of the students take calculus. Then it says "of these, 1/3 also take advanced chemistry." That means 1/3 of the calculus students also take advanced chemistry. Second thing: The word *least*. That means if possible answers include 11, 19, 127, 239, and 588, the correct answer is 11.

Okay, on to the approach. Algebra is our best friend here. I know you hate it, but it's not hard. We want the least number of students in the class, so let's call that x. Then:

$$\frac{x}{4} = \text{calculus students}$$

Remember, 1/3 of the calculus students also take chemistry, which can be expressed this way:

$$\frac{\frac{x}{4}}{3} = \text{calculus students who also take chemistry}$$

Now here's the tricky part. We know that 3 students take both courses, and there's a tendency to want to add the two values (x over 4 and x over 4 over 3). But think: it says 1/3 of the calculus students also take chemistry. So that's the number we've got: the number who take both courses:

$$\frac{\frac{x}{4}}{3} = 3 \longrightarrow \frac{x}{4} = 9$$

$$x = 36$$

| **Did you notice?** | Check your answer by plugging 36 into the original problem. If there are 36 students and 1/4 take calculus, that's 9 students. If 1/3 of those calculus students also take advanced chemistry, that's 3. Do you see that 1/3 times 1/4 equals 1/12, and that 1/12 of the students is a whole number (3)? That means the total number of students must be a multiple of 12. Check 12 and 24 to see that they don't work. |

**If they tell you the area of a square, you know the length.
And be on the lookout for 3-4-5 right triangles.**

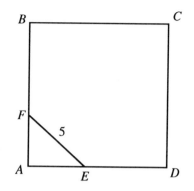

The area of square *ABCD*, above, is 100, and the
length of *BF* is 7. What is the length of *ED*?

Label the diagram as you go! When they tell you the area of a square, they're indirectly telling you the
length of each side of that square. The length is the square root of the area -- here the area is 100, so the
length is 10. And they've told us the length of *BF* is 7. Write it in!

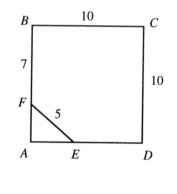

Now *AB* must be 10, because the thing's a
square. We know *BF* is 7, so *AF* must be 3.
And since right Δ*AEF* has a hypotenuse of 5
and a height of 3, we know it must be a 3-4-5
right triangle. So *AE* is 4.

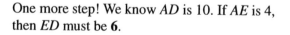

One more step! We know *AD* is 10. If *AE* is 4,
then *ED* must be **6**.

Did you notice?	They told us the length of *EF* in the diagram. The length of *BF* isn't indicated in the diagram but is revealed in the word problem. The length of the square is told indirectly, by giving you the area. And they don't tell you *AB* equals *AD*. You have to come to that conclusion on your own, based on what you know about squares. They give you all the information you need -- you have to put it together.

95 Make sure you understand what the chart or graph is telling you. When finding an average, always start with the formula and plug in what you know.

	Shaun	Christian
Game 1	10	8
Game 2	4	10
Game 3	10	9

The chart above shows the results of three basketball games between Shaun and Christian. What was the average (*arithmetic mean*) of the total number of points scored in each game?

This is a straight, regular old average question. What they're counting on is that a certain percentage of people get confused by charts, even simple ones. If you have any ability to read and understand charts, this question should take you about thirty seconds.

First, what are they telling us? The chart simply says that Shaun won the first game by a score of 10 to 8, and the third game by a score of 10 to 9. Christian won the second game, 10 to 4.

Second, what are they asking us? They want to know the average number of total points scored in each game. Let's begin with the formula for average, modified to match the details of this question:

$$A = \frac{\textbf{Total number of points}}{\textbf{Total number of games}}$$

The boys scored 18 points in Game 1, 14 in Game 2, and 19 in Game 3. That's a total of 51 points, so that will go at the top of the fraction. Next, they played 3 games, so that goes on the bottom, and we have this simple arithmetic problem to solve:

$$A = \frac{51}{3}$$

$$= 17$$

Did you notice?	We used "Total number of points" and "Total number of games" in the formula. Why? Because that's what they asked for. If they had asked for each boy's average number of points per game, we would have used 24 over 3 for Shaun and 27 over 3 for Christian. And if they had wanted the average number of points per boy over the three games, we would've used 51 over 2.

96 When parts of geometric shapes coincide, use those connections as stepping-stones to get to the answer.

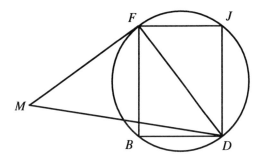

Rectangle *BFJD* is inscribed inside the circle above. If the circle has a circumference of 12π, what is the area of isosceles right triangle *MFD*?

This is similar to questions we've seen before about rectangles inscribed inside circles, and right triangles whose sides are also part of other polygons. We're given a clue that leads to another clue, that leads to another clue, that leads to the answer. So let's begin at the only logical place: with what we know.

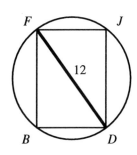

The circle has a circumference of 12π. We know the circumference of a circle is the diameter times π, so the diameter must be 12.

The diameter of the circle is also the diagonal of the inscribed rectangle (remember that?) The diagonal of the rectangle, then, is 12.

The diagonal of the rectangle is also side *FD* of the right triangle. So *FD* is 12.

An isosceles right triangle has two equal legs and a longer hypotenuse. Since the hypotenuse is obviously *MD*, side *MF* must be the other equal leg of the isosceles right triangle. Therefore, *MF* equals 12.

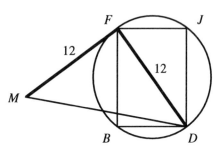

The area of a triangle is 1/2 (base x height). So:

A	**= 1/2 (12 x 12)**	*or*
	= 1/2 (144)	
	= 72	

A	**= 1/2 x 12 x 12**
	= 6 x 12
	= 72

Did you notice? The side of the triangle, the diagonal of the rectangle, and the diameter of the circle are all the same line. But each represents a different mathematical rule. Knowing those rules allows you to take the necessary steps to the answer. When multiplying 1/2 x base x height, it's often easier to multiply the 1/2 and one of the legs first. In the example above, 1/2 x 12 is 6. Then 6 x 12 is 72. Smaller numbers.

97

The probability of two successive events is the probability of the first times the probability of the second. But read carefully!

Hannah and Rebekah are the same size and have 3 pairs of shoes each: 1 white, 1 red, and 1 black. If all the shoes are put into a large bag, what is the probability of one of the girls pulling out, on the first try, a white pair she can wear?

There is major danger in this question, and it has nothing to do with fashion. But let's begin at the beginning.

Probability involves parts compared to wholes -- just like fractions, ratios, decimals, percents, and all that wonderful stuff. So we first need to figure out how many things we're dealing with. And that's danger number one.

Each girl has 3 pairs of shoes, so each girl has 6 shoes, and the two girls have a total of 12 shoes. So when we begin this little experiment, we have 12 shoes, and 4 of them are white. The probability, then, of pulling out a white shoe at random on the first try is 4 out of 12, or 1 out of 3. We can express this probability as a fraction if we want to (and I want to):

White shoe on the first try $= \dfrac{1}{3}$

This makes incredible sense, because for each girl, one-third of the pairs of shoes are white. Now pay attention. After one shoe has been chosen, we are dealing with a new situation. There are 11 shoes left and 3 of them are white. So the probability of choosing a white shoe on the second try (after getting one on the first try) is 3 out of 11.

BUT HOLD ON! The question says "a white pair she can wear." In order to wear these shoes, at least in public, there must be a right and a left. Did you think of that? So if the first shoe selected was a left shoe, we now need a right shoe to match it. And there aren't 3 right shoes -- there are only 2. The chances, then, of choosing a white shoe *she can wear* is 2 out of 11! That means the probability of choosing, on the first two picks, a white pair she can wear is:

$$\dfrac{1}{3} \cdot \dfrac{2}{11} = \dfrac{2}{33}$$

The answer to this question, and the answer you should grid in, is 2/33.

| **Did you notice?** | This question involves maybe a little bit more thinking than would be necessary on most SAT questions. More likely they'd be asking you about socks, and since socks should match in color but don't really come in left and right, you would only have had to deal with changing the probability on the second sock to 3 out of 11. But thinking a little extra now and then is okay. It's like swinging two bats. |

In a 30-60-90 triangle, the side opposite the 30-degree angle is *x*.
The hypotenuse is 2*x*, and the side opposite the 60-degree angle is √3*x*.

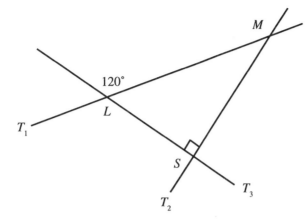

Lines T_1, T_2, and T_3 intersect to form $\triangle LMS$, as shown above. If MS equals $5\sqrt{3}$, what is the length of LS?

Although well-disguised, this question is really about 30-60-90 right triangles and the relationships among the lengths of their sides.

Do you see that $\angle LSM$ is 90 degrees? Okay, now look at that angle labeled 120 degrees. The one next to it, inside the triangle, is its supplementary angle: the sum of those two angles is 180. So the angle inside the triangle, $\angle MLS$, is 60 degrees.

We have a 90-degree angle and a 60-degree angle, so the third angle in the triangle, LMS, must be 30. (They have to add up to 180.) This is known as a 30-60-90 right triangle.

A 30-60-90 right triangle has several predictable characteristics. If you look at the first page of any SAT math section, you will see the following diagram:

It's saying that if the side opposite the 30-degree angle is *x*, then the hypotenuse (the side opposite the 90-degree angle) will be 2*x*. And the side opposite the 60-degree angle will be √3 times *x*.

So in a 30-60-90 triangle, if the side opposite the 30 is 10, then the hypotenuse is 20 and the side opposite the 60 is $10\sqrt{3}$.

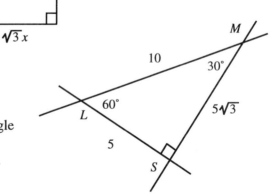

Back to the problem. The side opposite the 60-degree angle is $5\sqrt{3}$. That means the side opposite the 30-degree angle must be 5. So the answer to this question, at long last, is **5**.

Did you notice?	The diagram given with the formulas on the SAT has the 30-60-90 triangle lying on its back with the 90-degree angle at lower right. But when you encounter one in an actual question, the triangle may be oriented differently. In our example, you have to mentally rotate and flip one triangle to match the other. Once you do, you can see which side matches the *x*, which matches the √3*x*, and so on.

All circles contain 360 degrees. An arc is just some piece, or fraction, of that 360 degrees.

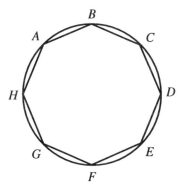

A regular octagon is inscribed inside a circle, as shown above. How many degrees does arc *DEFG* contain?

Circle? Degrees? Arc? What's the first thing that comes to your mind? (I mean after, "If I run away and join the Romanian National Guard, will I have to take math?") It should be the fact that a circle -- every circle -- has exactly 360 degrees. So the answer to this problem is going to be some fraction of that total. The question is, what fraction?

Well, what does the phrase *regular octagon* really mean? It means that the figure inscribed inside the circle has 8 equal sides. Further, the 8 angles inside that octagon are also equal. So what, you ask? I'll tell you so what. It just so happens that the regular octagon inscribed inside that circle is dividing the circumference of that circle into 8 equal arcs. Oh, now you're interested?

What do we have so far? The circle contains 360 degrees. It's been divided into 8 equal parts. One of those parts, then, must equal:

$$\frac{360}{8} \text{ or } \mathbf{45}.$$

Now watch where you jump. Each little section of that circle (for example, arc *AB*) contains 45 degrees. But that's not the answer to the question. They're asking for arc *DEFG*. Again, be careful. The arc we're interested in contains <u>three</u> of those 45-degree sections -- <u>not four</u>. So:

3 x 45 = 135.

And that's the answer (135).

A circle can be measured in three ways. First, it has area (the amount of space inside the circle). Second, it has a perimeter, called circumference, which is the length of the circle if it were a straight line. Third, it contains 360 degrees. An arc is a piece of the circle. It does not have area, but it has length (some fraction of the circumference) and it contains a certain number of degrees (some fraction of 360).

The ratios of corresponding sides of similar triangles are equal.

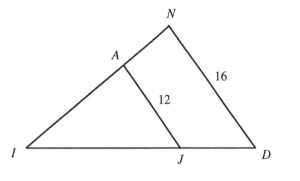

HMMM. THOSE TRIANGLES LOOK KIND OF SIMILAR.

In △*IND* above, *AJ* is parallel to *ND*, and *IN* equals 20. What is the length of *AN*?

It says way at the top of this page that the ratios of corresponding sides of similar triangles are equal. Well, what does *that* mean? It means that if two triangles are similar (their angles are equal), then the triangles are in proportion.

Look at the diagram. *AJ* and *ND* are corresponding sides. If you were to pull △*IAJ* out and sit it next to △*IND*, they would look like the same triangle, only one is a smaller version of the other. Side *AJ* would be the smaller version of, or correspond to, side *ND*. Do you see that *AJ* and *ND* have a ratio of 12/16, or 3/4? That ratio is the same for all three pairs of corresponding sides. So:

$$\frac{AJ}{ND} = \frac{IA}{IN} = \frac{IJ}{ID}$$

Each part of the smaller triangle is 3/4 the length of the corresponding part of the larger triangle. So:

$$\frac{IA}{IN} = \frac{3}{4}$$

And they told us that *IN* equals 20. Therefore:

$$\frac{IA}{20} = \frac{3}{4}$$

When we cross-multiply, we get 4(*IA*) = 60, and *IA* = 15. But don't put that as your answer, because it's wrong. Look again at the question. They're asking for *AN*, which is 20 - 15, or 5. The correct answer is **5**.

| **Did you notice?** | They don't tell you the triangles are similar. But by telling you those two lines are parallel, they're saying that the corresponding angles are equal. And since angle I is shared by both triangles, all three angles are equal, and the triangles are similar. That's how they tell you without telling you. Also notice, one more time, *where* the information is presented: a little here, a little there. You have to put it together. |

More Questions

No help this time, but if you need to wimp out, you'll find the answers at the end.

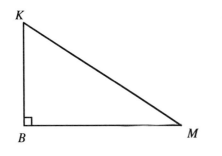

1. In $\triangle BKM$, $\angle M$ is 12 degrees less than $\angle K$. What is the measure of $\angle K$?

(A) 12 (B) 27 (C) 39 (D) 51 (E) 63

---

2. A woman travels 200 miles in 5 hours. If she continues at the same rate, how many miles will she travel in the next 3 hours?

(A) 15 (B) 80 (C) 120 (D) 320 (E) 600

---

3. Ann has $11. If the cost of greeting cards ranges from $1.50 to $3.50, what is the greatest number of cards she can buy?

(A) 7 (B) 6 (C) 5 (D) 4 (E) 3

---

4. If the product of two consecutive even integers is 48, what is their sum?

(A) 10 (B) 14 (C) 18 (D) 20 (E) 24

---

5. If $4a = 6b = 36$, the $a + b =$

(A) 15 (B) 18 (C) 20 (D) 24 (E) 30

6. If $7^2 = 4y + 3^2$, then $y =$

(A) 8 (B) 9 (C) 10 (D) 11 (E) 12

---

7. Which of the following has the least number of factors?

(A) 12 (B) 22 (C) 27 (D) 36 (E) 71

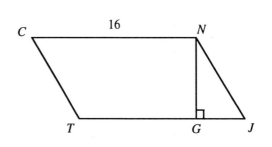

8. If the perimeter of parallelogram $TCNJ$ above is 52, and $JG = 6$, what is the area of the parallelogram?

(A) 10 (B) 60 (C) 96 (D) 128 (E) 160

---

9. The total cost of 8 identical flashlights is $104. What would the total cost of 6 of these flashlights be?

(A) $13 (B) $39 (C) $78 (D) $84 (E) $96

---

10. What is the volume of a cylinder if its height is 10 and the diameter of its base is 6?

(A) 4π (B) 16π (C) 36π (D) 64π (E) 90π

11. A bag contains 40 marbles: 15 blue, 10 red, 9 green, and 6 yellow. What is the probability that, on the first random selection, either a red or yellow marble will be drawn?

(A) 1/4   (B) 2/5   (C) 3/20 (D) 3/5 (E) 3/4

12. For all integers x, let
$[x] = x (x - 1)$.
What is the value of $[5] + [2]$?

(A) 2   (B) 7   (C) 20   (D) 22   (E) 42

13. The three angles of a triangle are in the ratio 2:3:4. What is the measure of the <u>smallest</u> angle?

(A) 20   (B) 40   (C) 60   (D) 80   (E) 90

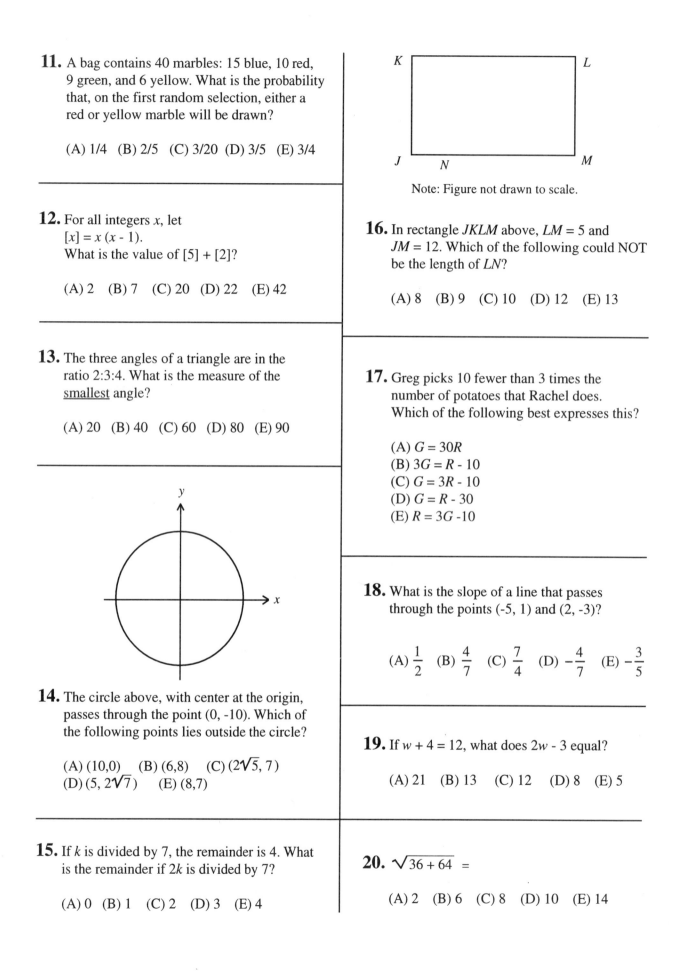

14. The circle above, with center at the origin, passes through the point (0, -10). Which of the following points lies outside the circle?

(A) (10,0)   (B) (6,8)   (C) $(2\sqrt{5}, 7)$
(D) $(5, 2\sqrt{7})$ (E) (8,7)

15. If k is divided by 7, the remainder is 4. What is the remainder if $2k$ is divided by 7?

(A) 0   (B) 1   (C) 2   (D) 3   (E) 4

Note: Figure not drawn to scale.

16. In rectangle $JKLM$ above, $LM = 5$ and $JM = 12$. Which of the following could NOT be the length of LN?

(A) 8   (B) 9   (C) 10   (D) 12   (E) 13

17. Greg picks 10 fewer than 3 times the number of potatoes that Rachel does. Which of the following best expresses this?

(A) $G = 30R$
(B) $3G = R - 10$
(C) $G = 3R - 10$
(D) $G = R - 30$
(E) $R = 3G - 10$

18. What is the slope of a line that passes through the points (-5, 1) and (2, -3)?

(A) $\dfrac{1}{2}$ (B) $\dfrac{4}{7}$ (C) $\dfrac{7}{4}$ (D) $-\dfrac{4}{7}$ (E) $-\dfrac{3}{5}$

19. If $w + 4 = 12$, what does $2w - 3$ equal?

(A) 21   (B) 13   (C) 12   (D) 8   (E) 5

20. $\sqrt{36 + 64} =$

(A) 2   (B) 6   (C) 8   (D) 10   (E) 14

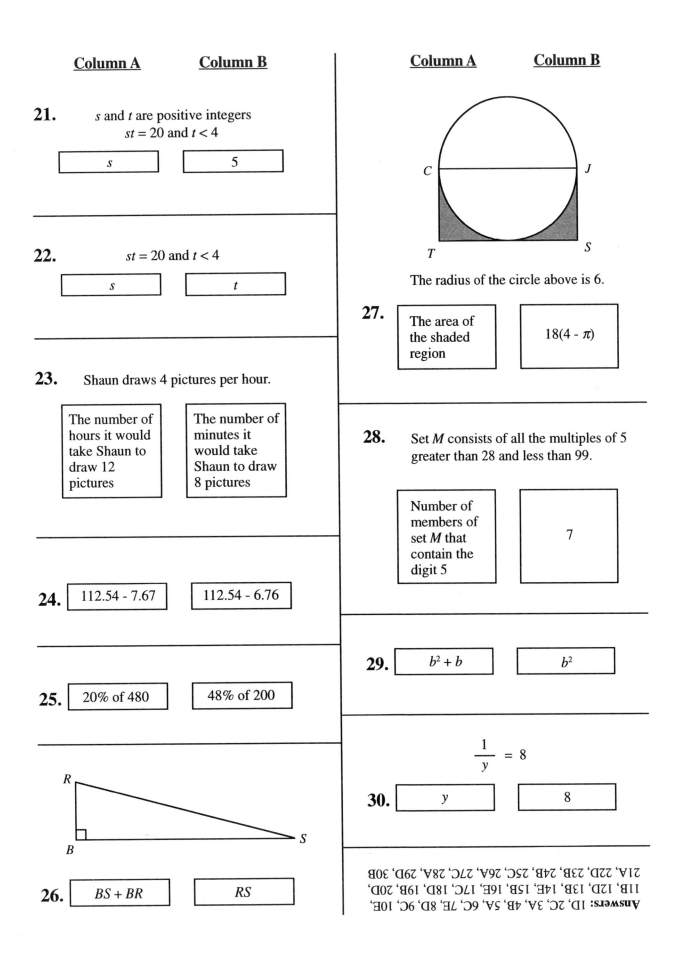

Column A	**Column B**

21. s and t are positive integers
$st = 20$ and $t < 4$

s	5

22. $st = 20$ and $t < 4$

s	t

23. Shaun draws 4 pictures per hour.

The number of hours it would take Shaun to draw 12 pictures	The number of minutes it would take Shaun to draw 8 pictures

24.

112.54 - 7.67	112.54 - 6.76

25.

20% of 480	48% of 200

26.

$BS + BR$	RS

Column A	**Column B**

The radius of the circle above is 6.

27.

The area of the shaded region	$18(4 - \pi)$

28. Set M consists of all the multiples of 5 greater than 28 and less than 99.

Number of members of set M that contain the digit 5	7

29.

$b^2 + b$	b^2

$$\frac{1}{y} = 8$$

30.

y	8

Answers: 1D, 2C, 3A, 4B, 5A, 6C, 7E, 8D, 9C, 10E, 11B, 12D, 13B, 14E, 15B, 16E, 17C, 18D, 19B, 20D, 21A, 22D, 23B, 24B, 25C, 26A, 27C, 28A, 29D, 30B

Glossary

Acute Angle   An angle that contains less than 90 degrees.

Adjacent Angles   Two angles sharing one common side and a common vertex.

Angle   Formed by two lines that either intersect (cross each other) or start from the same point. The opening of the angle is measured in degrees and can be thought of as a fraction of a circle's 360 degrees.

Arc   A section of a circle, also known as a little curved thing.

Area   The amount of space inside a two-dimensional (flat) object. Area is measured in square inches, square centimeters, square miles, and so on.

Average   The number you get when you add a group of numbers and then divide by how many there are. The average of 6, 15, and 12 is 11. Also called *arithmetic mean*.

Bisect   To cut exactly in half. A line that bisects a 70-degree angle divides it into two 35-degree angles.

Chord   A line connecting one point on a circle with another.

Circle   A collection of all the points that are the same distance from one, central point.

Circumference   The "length" of a circle, measured from one point and then all the way around back to that point. It's really the perimeter of the circle, if you know what that is. If you don't, look it up right now.

Complementary Angles   Two angles containing a total of 90 degrees.

Congruent Triangles   Triangles that have the same angles and are the same size (the three angles and three sides of one triangle are equal to the three angles and three sides of the other).

Consecutive   One after the other, without skipping. The consecutive positive even integers are: 2, 4, 6, 8... The consecutive positive odd integers are: 1, 3, 5, 7... The first five consecutive positive integers are: 1, 2, 3, 4, and 5. The series -2, -1, 0, 1, 2... contains consecutive integers.

Cross Multiply   With two fractions expressed as an equality, to multiply the numerator of one fraction times the denominator of the other, and the denominator of the first fraction times the numerator of the other.

Cube   A rectangular solid with six square sides, or faces. Also, a number raised to the 3rd power (8 is the cube of 2).

Cylinder   A solid whose base and top are identical parallel circles. (In other words, anything shaped like a pipe or a tube or a can of chicken & stars soup.) On the SAT, the sides of a cylinder are always perpendicular to the bases, unless otherwise stated.

Degree	A unit of measure used to describe the size of an angle or a part of a circle. A complete circle has 360 degrees. A thermometer also has degrees, but what the connection is, who knows?
Denominator	The bottom number of a fraction. The denominator of 2/5 is 5.
Diagonal	On the SAT, the line connecting the vertices of two opposite angles of a quadrilateral. Also, the length of that line.
Diameter	The longest line you can draw from one point in a circle to another (the longest chord). The diameter is two times the length of the radius.
Difference	The number you get when you subtract one number from another.
Digit	A whole numeral. There are ten digits: 0, 1, 2, 3, 4, 5, 6, 7, 8, and 9. The whole number 6,300 contains four digits.
Doughnut	Any ring-shaped cake that is made of either dough or batter and cooked in hot oil. (You may be asking why I needed to include this. Well, why did you need to look it up?)
Equation	An arithmetic or algebraic expression that has some value on the left and some value on the right side of an equal sign.
Equilateral	Having all sides equal. An equilateral triangle has three equal sides.
Exponent	A little number placed next to a big number. Whatever the little number (the exponent) is, the big number is multiplied times itself that many times. So $4^3 = 4 \times 4 \times 4$. (The numeral 3 is the exponent.)
Factor	Numbers that are multiplied together. The whole number factors of 6 are 1, 2, 3, and 6, because $1 \times 6 = 6$ and $2 \times 3 = 6$.
Glossary	A list of words (usually the ones you already know) and what they mean.
Height	The line perpendicular to the base of a polygon, or the length of that line. To find the height of things like triangles and parallelograms, draw a line from the highest point straight down to the base. In a right triangle, one of the legs is also the height. In a parallelogram, the side is not the height.
Hence	"From now." They use it in word problems sometimes: "...How old will Matt be 7 years hence?" How old will he be 7 years from now?
Hexagon	A 6-sided polygon.
Hypotenuse	In a right triangle, the longest side (the side opposite the right angle).
Inclusive	Including the two extremes. The integers from 0 to 5, inclusive, are: 0, 1, 2, 3, 4, and 5.
Integer	A whole number that is either positive, negative, or zero. These are integers: -5, 0, 2, 308. These are not: 3/4, 1.6, 19.4. (There is a complex system for classifying numbers in mathematics. For SAT purposes, however, an integer is a whole number and a whole number is an integer.

Isosceles Triangle	A triangle having at least two equal sides and two equal angles.
Median	The middle number in a group. So in the following bunch -- 4, 5, 7, 10, 13, 18, and 30, the median is 10.
Midpoint	Just what it sounds like: the point equidistant (or equally-distant) from the ends of a line. So B is the midpoint of AC if AB equals BC.
Multiple	The product of two or more whole numbers. For example, 40 is a multiple of 5, because 5 x 8 = 40. It is therefore also a multiple of 8.
Numerator	The top number of a fraction. The numerator of 2/5 is 2.
Obtuse Angle	An angle whose measure is more than 90 and less than 180 degrees.
Origin	The place where the x-axis crosses the y-axis (usually referred to as the point (0,0).
Parallel Lines	Lines in the same plane that never meet.
Parallelogram	A quadrilateral whose opposite sides and opposite angles are equal.
Pentagon	A 5-sided polygon. Also, a really big building in Washington.
Perimeter	The sum of the lengths of all the sides of a polygon.
Perpendicular	Two lines meeting to form right angles. A stop sign is perpendicular to the road, except right after a hurricane or a student driver has passed by.
Pi	A Greek letter (π) used to represent the relationship between the radius of a circle and its circumference and area. (Circumference = $2\pi r$. Area = πr^2.) Taken to five decimal places, $\pi = 3.14159$.
Prime Number	A number whose only positive whole number factors are 1 and itself. So 7 is a prime number because other than 1 x 7, there is no other way to multiply two positive whole numbers and get 7.
Probability	The mathematical chance of something happening. Probability can be expressed as a fraction, ratio, decimal, or percent. So the probability of flipping a coin and getting tails is 1/2, or 1:2, or 0.5, or 50%.
Product	The number you get when you multiply two numbers.
Pythagorean Theorem	Formula used to describe the relationship among the lengths of the sides of a right triangle ($a^2 + b^2 = c^2$).
Quadrilateral	A four-sided polygon. The angles of a quadrilateral total 360 degrees. All rectangles are quadrilaterals, but not all quadrilaterals are rectangles.
Quotient	The number you get when you divide one number into another.

Radical	The symbol ($\sqrt{\ }$) that tells you to find the square root of a number.
Radius	The distance from the center of a circle to any point on the circle. The plural of radius is radii.
Ratio	Comparison of two or more numbers, expressed in lowest terms. So the ratio of three angles (20°, 40°, and 80°) is 1:2:4.
Rectangle	A quadrilateral that has four 90-degree angles.
Remainder	The number left over after you've divided one number into another.
Respectively	"In the same order." Meaghan, Jenna, and Ryan are 12, 9, and 5, respectively. That means Meaghan is 12, Jenna is 9, and Ryan is 5.
Right Triangle	A triangle that contains a 90-degree angle.
Semicircle	A half-circle.
Similar Triangles	Two triangles with exactly the same three angles. Similar triangles do not have to be the same size.
Slope	The "steepness" of a line. Determine the slope with a fraction: the change in vertical measure is on top, the change in horizontal measure is on bottom.
Square	A quadrilateral that has four 90-degree angles and four equal sides. Also, a number raised to the 2nd power (9 is the square of 3).
Square Root	One of two equal factors of a number. The square root of 36 is 6, because 6 times 6 is 36.
Sum	The number you get when you add two or more numbers.
Supplementary Angles	Two angles containing a total of 180 degrees.
Triangle	A three-sided polygon. The angles of a triangle total 180 degrees.
Unit's Digit	The number all the way on the right. For example, in the number 11,763, the 3 is the unit's digit. (In this example, 6 is the ten's digit, 7 the hundred's digit, etc.)
Variable	A number or group of numbers represented by a letter or symbol. The value of the variable can change (vary) depending on the mathematical relationship defined for it.
Vertex	The point where two lines meet to form part of a polygon (such as the corners of a triangle. The plural of *vertex* is *vertices*.
Volume	The amount of space contained in, or occupied by, a solid figure. Volume may be expressed in cubic feet, cubic centimeters, cubic miles, etc. Volume also refers to the loudness of something, as in, "Turn down the volume on that TV before I throw it into the canal!"
Whole Number	See *Integer*.

Index

This index provides a list of SAT math topics, followed by the questions in which those topics are addressed. Numbers in bold refer to topics appearing in the More Questions *section.*

500 SAT® WORDS AND HOW TO REMEMBER THEM FOREVER!

This book does more than just teach you which words to learn. It teaches you <u>how</u> to learn them -- and not for temporary memorization, but permanent retention. Using visual images, *500 SAT Words* helps you build a mental bridge between each unfamiliar word and its simple meaning. It also points out look-alike and sound-alike words frequently used as traps on the SAT, provides spelling, correct pronunciation, and alternate parts of speech for each word, and gives sample sentences for many of the entries. Humorous cartoons and a list of 135 additional words add to the effectiveness of this powerful vocabulary-building tool. Great for anyone who wants to develop a strong command of the English language. *(120 pages, $14.95)*

If your local educational resource store doesn't have *500 SAT Words*, ask them to order it today!

Mostly Bright Ideas
888-301-2829
http://www.mostlybrightideas.com

Burdick Associates
800-828-1434
716-335-8807 (fax)

EXTOL (ex-TOLE) *verb* — **praise**

Sounds like: eggs toll

Picture: A giant egg working as a toll collector. The driver must praise the egg before he's permitted to pass through the gate.

AVARICE (AH-ver-iss) *noun* — **greed**

Looks like: Ava rice

Picture: A woman named Ava is seated at a table with her arms around a mound of rice. She won't share the rice with anyone. She's greedy.

BALEFUL (BAIL-full) *adj* -- **harmful; menacing**

Sounds like: bale fall

Picture: Gigantic, heavy of hay falling from the

UPBRAID (up-BRA -- **criticize seve**

Looks like: "up brai

Picture: A girl with b She's being scolded s by her teacher that her are standing straight u